Advance Praise for *Chaco Astronomy*

"The Solstice Project papers are a kind of mythic journey. From serendipitous discovery to substantiated insight, they chronicle the birth and growth of completely new perspectives on the fundamental mystery of the Chaco phenomenon. If you've mused around a campfire, dreamed into a star-strewn sky, or stood dumb in the aching silence of the high desert, you know how the human imagination yearns to roam. This book harnesses that passion to the rigor of observation, measurement, and analysis, and delivers a grounded entrée into a world that will always resist complete understanding."

— **Adriel Heisey**, photographer and author *Under the Sun: A Sonoran Desert Odyssey*

"Anna Sofaer's pioneering work on Chaco Canyon, a World Heritage Site, should be required reading for anyone interested in how the prehistoric people of the American Southwest conceptualized their universe and placed themselves within that universe."

— **Todd W. Bostwick**, co-editor *Viewing the Sky Through Past and Present Cultures:*
Selected Papers from the Oxford VII International Conference on Archaeoastronomy

"Here, gathered for the first time in one accessible volume, is the research on the famous Anasazi 'Sun Dagger' and other astronomical evidence from Chaco Canyon. Anna Sofaer's exciting, painstaking and controversial work remains essential for anyone fascinated by the skywatching traditions of the Southwest."

— **Evan Hadingham**, author *Early Man and the Cosmos*, and producer of PBS's *NOVA*

"The Solstice Project's work has been controversial because it reveals the extraordinary scientific knowledge and technical abilities possessed by the builders of the Chacoan Big Houses. It has forever changed both the public's and the professionals' understanding."

— **Flora Clancy**, author *Sculpture in the Ancient Maya Plaza: The Early Classic Period*, Professor Emeritus
of Art History, University of New Mexico

"The Solstice Project's in-depth research of the Sun Dagger phenomenon brought self-discovery for many Pueblo educators. The question of self, the sense of place, the roots in relation to the place labeled Chaco Canyon further heightened our awareness of who, where, and what we as Pueblo people were encountering."

— **Kirby Gchachu**, Zuñi astronomer, Southwest Indian Polytechnic Institute

"Thirty years of research in the field of archaeoastronomy by Anna Sofaer and her Solstice Project colleagues have broadened the way archaeologists and the public think about ancestral Pueblo culture in the American Southwest. *Chaco Astronomy: An Ancient American Cosmology*, which brings the fruits of this research together for the first time as a single book, will be invaluable to scholars and students as well as general readers with a special interest in the topic."

— **David Grant Noble**, editor *In Search of Chaco and New Light on Chaco Canyon*

"Research conducted by the Solstice Project has demonstrated that the Chacoans manipulated their physical environment—using rock art, architecture, and landscape features—to express astronomical and cosmological relationships. This collection of nine papers (previously published in scattered venues) brings together the Solstice Project's seminal work in one convenient package. Anna Sofaer's introduction provides an excellent historical background plus an overview of the project's findings."
 — **Ruth Van Dyke**, author *Experiencing Chaco: Landscape and Ideology at the Center Place*

"These studies have appeared in specialized journals and books and only intermittently over nearly 30 years. Streamlining access to her most substantive publications, Anna Sofaer has added a singular volume of archaeoastronomical inquiry, which will shed light on ancient Chaco culture for many solstices to come."
 — **E. C. Krupp**, Director, Griffith Observatory, author *Echoes of the Ancient Skies: The Astronomy of Lost Civilizations* and *In Search of Ancient Astronomies*

"Over the last three decades, works by Anna Sofaer and her colleagues in the Solstice Project have richly stimulated thinking about astronomy in indigenous societies of the ancient Southwest. Bringing their diverse publications together in one volume is an invaluable contribution, greatly enhancing opportunities for scholarship and learning, with thought-provoking implications reaching far beyond the specific areas of their research."
 — **Wendy Ashmore**, co-editor *Archaeologies of Landscape: Contemporary Perspectives*, Professor of Anthropology, University of California, Riverside

"This work, based on empirical facts and collaborative work with colleagues from astrophysics, geodesy, archaeology, and anthropology, has great precision in the observation and measurement of astronomical phenomena of this extraordinary site. Sofaer's research has made an important contribution to evaluating Chaco's sacred landscape, astronomy, architecture and geometry in relation to the ancient civilization of Mesoamerica and its northern transitional regions. Another important contribution is the Solstice Project's joint research with Hopi, Acoma and other Pueblos as well as Navajo communities regarding the cosmological concepts of these living cultural traditions and their relation to Chaco."
 — **Johanna Broda**, co-author *The Great Temple of Tenochtitlan: Center and Periphery in the Aztec World*, Ethnohistorian, National Autonomous University of Mexico (UNAM)

"After over 25 years of careful observation, consultation with tribal representatives of the descendant communities, and a process of painstaking documentation, Anna Sofaer and the Solstice Project have reached yet another important milestone with the publication of *Chaco Astronomy*. This new volume further solidifies theory-based observations that are rooted in traditional indigenous knowledge and in the meaningful and practical study by the Solstice Project of the natural, spiritual, and complex composition of this resource that will remain significant to its descendants and to the many who will become exposed to the grandeur and mystery of such an ancient canyon."
 — **Brian D. Vallo**, Acoma Pueblo, Founding Director, Sky City Cultural Center and Haaku Museum

"The work of Anna Sofaer concerns more than Chaco Canyon or the astronomy of ancient America. Her discoveries suggest that human beings know more than we think we know, and sense more than we believe—not only within the realms of smell, taste, touch, sound, but in how we understand our place in the universe. The timeless message from Chaco is this: a deeper connection to the natural world offers us a larger awareness of life and death, and with that awareness, humility."
 — **Richard Louv**, author *Last Child in the Woods: Saving Our Children from Nature Deficit Disorder*

CHACO ASTRONOMY

AN ANCIENT AMERICAN COSMOLOGY

Anna Sofaer

and Contributors to the Solstice Project

OCEAN TREE BOOKS

Santa Fe, New Mexico

OCEAN TREE BOOKS
Post Office Box 1295
Santa Fe, New Mexico 87504
www.oceantreebooks.com
(505) 983-1412

Papers in this volume originally appeared in the following books or periodicals and are reprinted here by expressed permission of the original publishers or institutions: *Science; Archaeoastronomy in the New World* (Cambridge University Press); *Astronomy and Ceremony in the Prehistoric Southwest* (Maxwell Museum of Anthropology, University of New Mexico Press); *Celestial Seasonings: Astronomical Connotations of Rock Art in the Prehistoric Southwest (1994 International Rock Art Congress Proceedings,* American Rock Art Research Association); *Anasazi Architecture and American Design* (University of New Mexico Press); *Architecture of Chaco Canyon, New Mexico* (University of Utah Press); *Time and Astronomy at the Meeting of Two Worlds* (Warsaw University); *World Archaeoastronomy* (Cambridge University Press); *Archaeoastronomy; Acts of History: Ritual, Landscape and Historical Archaeology* (University Press of Colorado); *Bulletin of the American Astronomical Society;* and the Society for American Archaeology.

SOLSTICE PROJECT
www.solsticeproject.org

Design by Richard Polese. Production by Sunflower Designs, Santa Fe.

Cover photograph: *Summer Solstice Sun Dagger, Fajada Butte, Chaco Canyon,* by Karl Kernberger (copyright © Solstice Project).

Pages 21, 79, 127, 143, 167: *Rock Slabs of the Sun Dagger Site* viewed from cliff face, photograph by Karl Kernberger (copyright © Solstice Project).

ISBN: 978-0-943734-46-0

Library of Congress Cataloging in Publication Data:

Sofaer, Anna.
Chaco astronomy : an ancient American cosmology / by Anna Sofaer and contributors to the Solstice Project.
        p. cm.
Includes bibliographical references.
ISBN 978-0-943734-46-0 (alk. paper)
1. Chaco astronomy. 2. Chaco cosmology. 3. Chaco Culture National Historical Park (N.M.) —Antiquities. I. Title.
E99.C37S65 2007
978.9'8201--dc22
                                                                                     2007036916

For Mike

Contents

Light Markings and Petrogylphs on Fajada Butte

Astronomical Expressions in the Major Chacoan Buildings

The Great North Road

Computer Restoration of the Sun Dagger Site

Appendix: Abstracts

Sun setting through doorway at Pueblo Pintado, an outlying Chaco building, photographed by Anna Sofaer. (Copyright © Solstice Project.)

Foreword

IN THE EARLY 1980S, I ATTENDED a meeting of the American Indian Science and Engineering Society in Los Angeles. As part of the program, Anna Sofaer and Rolf Sinclair showed her documentary, *The Sun Dagger*. It profoundly affected me. If what Sofaer and Sinclair proposed were true, it offered all of us a new paradigm for understanding Pueblo history. The paradigm acknowledged the ancestral Pueblo people's worldview. It theorized that ancient Puebloans had integrated long-term scientific observation, art in the form of shadow and light, and spirituality into a singular and complex construction.

When Anna spoke, she asked for questions. As I listened, I sensed that the project's discoveries had met with significant resistance within the academic community.

After her talk, I introduced myself to Anna and suggested that I might be of assistance to the Solstice Project. As Deputy Director of the National Geodetic Survey (NGS), I explained to Anna that my agency was responsible for performing all of the precise geodetic surveys for the civilian side of the government. We had the capability to document accurately the position and orientation of any structure relevant to her research. Moreover, if we did perform a survey, we had final authority in certifying its accuracy.

I wanted to help the Solstice Project for several reasons. First, if Anna's theories proved true, I sensed their immense value for Native Americans, especially younger ones, in revealing the scientific achievements of their ancestors. Second, I wanted to confront the assumption by some that ancestral Pueblo people were incapable of creating architectural constructs combining lunar and solar alignments. Finally, this offered a unique opportunity to use geodetic techniques to help document a Southwestern archaeological site, something that no one, to the best of my knowledge, had ever done.

Thus I arranged for Anna to come to the National Geodetic Survey's headquarters in Rockville, Maryland to brief our Director, Capt. John D.

Bossler, about the Solstice Project. Her research impressed him. Shortly after this meeting, we reached an agreement: The National Geodetic Survey would document the geodetic azimuth (the orientation in relationship to true north) of all the major buildings in Chaco Canyon.

We started the surveys in September 1984. Because the Global Positioning System (GPS) was not a common utility at the time, we used the classic geodetic technique of observing the star Polaris to determine each azimuth. Today, geodesists rarely observe an astronomic azimuth; however, for short lines, this technique is more accurate than GPS.

Once we had a reference azimuth running parallel to a main building wall, accurate to plus or minus three seconds of arc, we measured offsets from the wall at one meter spacing and calculated the azimuth of the wall using the "least squares" adjustment technique. This averaging technique allowed us to compensate for any irregularities in the walls. By using this method, we could be certain that our measurements represented the original wall alignments.

Analysis of the original 1984 surveys and subsequent surveys showed that twelve of the fourteen major buildings in Chaco Canyon align to either the sun or the moon. Further analyses demonstrated inter-building lunar alignments over distances up to 26.7 km. We found lunar alignments between three sets of buildings:

- Pueblo Pintado—Chetro Ketl—Kin Bineola
- Kin Bineola—Peñasco Blanco—Una Vida
- Peñasco Blanco—Pueblo Bonito—Una Vida.[1]

I wish I could say that the Solstice Project's findings and interpretations of the buildings' solar and lunar alignments found wide acceptance in the archaeological community. That did not happen. Trying to understand the underlying causes for this runs a gauntlet of psychological and intellectual obstructions.

The Southwestern archaeological community

is generally conservative. This is not necessarily negative: In a proper context, acceptance of change occurs after careful examination of evidence and analysis of merit. Often a new concept must accrue supporting confirmation from other investigators before it matures to become part of accepted doctrine. Unfortunately other factors can color the acceptance process, such as gender or age bias, racial prejudice, academic status, and narcissism, along with its accompanying envy. Logically few people would admit to or subscribe to this litany of attributes as reasons to reject a new idea.

Let us explore academic status. The rationale for having formal education and training in the discipline in which you engage is reasonable. However, that alone can be a rigid requirement for acceptance of a theory. What we forget regarding the field of southwestern anthropology and the related fields of archaeology and archaeoastronomy is their short history of less than one hundred years.

Many early scholars, now revered for their contributions to archaeology, began their careers without proper credentials, as judged by today's criteria. We forget the circuitous routes followed by some of the archaeological pioneers: Adolph Bandelier began his studies in law; Edgar L. Hewett was superintendent of schools in Florence, Colorado; A.V. Kidder started his studies at Harvard as a premedical student; J. Walter Fewkes began his career at Harvard as a zoologist. In addition, A. E. Douglass was an astronomer.

Institutions and professions create their culture through their attitudes and behavior patterns. Like any culture, what is habitual can be incorporated as a value. Periodically, something happens that causes the establishment to examine critically its inner workings. One such event was the passage in 1990 of the Native American Graves Protection and Repatriation Act (NAGPRA). This legislation required institutions or researchers receiving federal funding to consult with affected Native Americans on the disposition and repatriation of human remains and cultural artifacts. It also vested Native Americans with the right to determine final disposition of their ancestors'

---

1. Sofaer, Anna, *The Primary Architecture of the Chacoan Culture: A Cosmological Expression*, Solstice Project, Washington, D.C., 1997.

remains and grave goods. The overall result was improved relationships and communications between Native Americans and museums, institutions and individuals.

Sometimes a person from within the community issues a wake-up call. For example, Walter W. Taylor, in his 1948 publication, *A Study of Archaeology,* cautioned archaeologists that they were only engaging in "historical reconstruction" if they did not pay attention to cultural processes. This admonition stung some members of the community. However, his criticisms led to the benefits of a holistically multi-disciplinary approach in studying the past.

We have reached a time in our technologically driven society when no one person has the capacity to be proficient in more than one or two of the social, physical, computer or biological sciences. I do not advocate an uncritical acceptance of all who wander into the fold with a new idea or concept. However, we need to think inclusively rather than exclusively.

When I asked Anna, more than twenty-five years ago, if I could help her, I felt I could bring something of value to the Solstice Project. As I have worked with her over these many years, her approach has impressed me: she never asked inappropriate questions about Pueblo beliefs. She never interpreted our cultural or spiritual way of life. In other words, she kept her focus on the scientific investigations of the Chacoan sites without intruding into sensitive areas. Pueblo people respect and trust her. Furthermore, like her early archaeological predecessors, she has become an expert in her field, that of archaeoastronomy. Finally, she has three qualities that continue to serve her well: indomitable courage, unwavering persistence and remarkable intuition.

I understand some of the initial skepticism that met the Solstice Project when it first tendered its theory that the ancestral Pueblo people had incorporated solar and lunar alignments in the stone construction atop Fajada Butte. However, this book, *Chaco Astronomy: An Ancient American Cosmology,* demonstrates clearly that the Chacoans redundantly marked the solar and lunar cycles by constructing large-scale buildings and creating light markings on small petroglyphs. Awareness of such lunar alignments reached a tipping point in 1988 with J. McKim Malville's independent finding that an outlying Chacoan building is also aligned to the lunar major standstill ("Lunar Standstills at Chimney Rock" in *Archaeoastronomy,* supplement to *Journal for the History of Astronomy,* 16). Recent research in Mexico, by archaeologists Alonso Mendez and Christopher Powell, indicates the existence of other lunar alignments.

I imagine a few diehards, whatever their reasoning, will never accept the idea that ancestral Pueblo people commemorated the lunar cycle through monumental constructions. I encourage those skeptics to spend a few hours with the Solstice Project's interactive computer model of the Sun Dagger. If this remarkable achievement, a result of over twenty-five years of inter-disciplinary work by astronomers, archaeastronomers, archaeologists, computer modelers, mathematicians, geodesists, graphic artists and photogrammetrists, does not convince them, then I can only respect the faith of their convictions.

Phillip Tuwaletstiwa
Galisteo, New Mexico, June 2007

Phillip Tuwaletstiwa holds degrees in geodetic science from Ohio State University and a Master of Science degree from Cornell's School of Civil Engineering. For over two decades, he served with the National Oceanic and Atmospheric Administration in several positions, including deputy director of the National Geodetic Survey. He managed an applied GPS research program at Ohio State and later developed a comprehensive Land Information System for the Hopi tribe, which mapped over 400 Hopi sacred sites. He serves on the board of the Indigenous Language Institute in Santa Fe and the Hopi Wildlife Endowment Fund.

Fajada Butte from the back wall of Hungo Pavi, Chaco Canyon, photographed by John Butterfield. (Copyright © John Butterfield.)

Introduction

Anna Sofaer

A N AUSPICIOUS OBSERVATION high on a butte in Chaco Canyon in the extreme heat of late June 1977 initiated a study of a previously unrecognized astronomy of ancient America. A deeper understanding of this ancient science unfolded over the next three decades, through the Solstice Project's collaboration with numerous scholars and scientists, to reveal a civilization with a complex cosmology.

In this first collection of the papers of the Solstice Project, we present our findings on the astronomy and cosmology of the ancient Chaco culture of New Mexico. This book brings forth thirty years of our published and soon-to-be-published research and documentation. The papers reveal that the people of Chaco, a culture that thrived in the arid San Juan Basin from A.D. 850 to 1130, developed an elaborate commemoration of the cycles of the sun and the moon. The Chaco people expressed this commemoration through the alignments of magnificent multistoried buildings, through the orientation of the Great North Road, and by numerous seasonal sunlight and lunar shadow markings on petroglyphs.

Our understanding of the complexity of the Chaco culture's astronomy deepened as, year to year, we developed new observations and insights. With our nine research papers in hand, many of which are currently out of print or hard to find, readers can now appreciate the interrelationships of these findings. Consistently, it was our experience that each set of data that we developed—while it would be independent of the previous data sets and stand on its own—remarkably corresponded with and reinforced our earlier results. For instance, the Chaco people's solar and lunar light markings on petroglyphs expressed an engagement with the sun and moon that we would later find in their architecture in three other separate expressions: in the solar-lunar expressions of both the alignments and the geometries of the Chacoan buildings, and in the astronomical regional pattern formed by their interbuilding relationships.

THE SOLSTICE PROJECT

In 1977, I was recording petroglyphs as a participant in a field school, a ten year project that had been organized by the Archaeological Society of New Mexico to create an archival record of the rock art of Chaco Canyon. My coworker, Jay Crotty, a skilled climber and surveyor, and I were assigned Fajada Butte, a prominent feature rising 400 feet from the canyon floor. As we climbed the butte on June 28, we recorded more than twenty petroglyphs. Late in the afternoon, when we reached the top, we saw a large spiral petroglyph carved on the cliff face behind three large rock slabs. The spiral was deeply shadowed so we decided to return the next morning when there would be more light for our photographing. We happened to reach the site close to noontime, eight days from summer solstice, at the very moment that a "dagger" of light was bisecting the spiral.

Three months earlier I had watched the shape of a serpent's shadow form on the Maya pyramid of Chichen Itza as the sun set at equinox[1]; and, only one month earlier, I had seen sequential photographs taken by an anthropologist of a pictograph of a human figure in a Baja California cave. The series showed a horizontal dart of light bisecting the figure's eyes as the sun rose at winter solstice.[2]

These experiences, coupled with my studies of Maya and other ancient astronomies over the past few years, helped me to recognize that the event I witnessed on Fajada Butte was a marking of the summer solstice. This light dagger, as we spontaneously described it in our notes, descending through the central position on the spiral, seemed undeniably to mark the highest position of the sun in the year and possibly in the day.

I soon learned that this was the first time such an event had been observed in Southwestern studies. A few researchers had sensed some astronomical expression in Chaco and they had surveyed precise cardinal orientations in two central Chaco buildings.[3] Yet, at the time we observed the dagger of light on the Fajada spiral, there had been little attention paid to Chaco's cosmology. This absence of interest may have been in part because the Chaco people, unlike their neighbors to the south, left no written record that might offer clues to a possible cosmology.

After the experience of finding the Sun Dagger, many questions presented themselves: Did it mark other times of the year and solar noon? If so, how precisely? How did the rock slabs form the light and shadow patterns? For what purpose did people create it?

In the spring of 1978, I organized the nonprofit Solstice Project. Our objectives were three-fold: First, we set out to study and preserve the Sun Dagger site and other astronomical expressions of the Chaco culture. Second, we hoped to bring out our findings to the growing field of archaeoastronomy. Finally, we planned through educational programs to convey our research to the interested public. Over the next three decades the intricacy of this site, and the many other constructions that we ultimately studied in Chaco, would call upon the highly developed skills of many individuals from different fields.

Inspired by our early evidence of solar markings on Fajada Butte, several critical helpers joined our study: Volker Zinser, an architect trained in Germany in shadow and light formation, astutely analyzed the Sun Dagger site's geometry of shadow configurations. Rolf Sinclair, an experimental physicist, applied his rigorous analytical skills and conducted experiments to test our theories of the site. Karl Kernberger, noted for his painstaking photography of subtle rock art sites, helped us develop a precisely timed record of the site's light markings throughout the solar cycle of 1978. A year later, LeRoy Doggett, an astronomer with the U.S. Naval Observatory, shared his extensive understanding of the moon's 18.6 year standstill cycle, just as we discovered at the Sun Dagger site shadow patterns that appeared to mark the extremes of this cycle.[4] A number of archaeologists, including Chaco scholar Gwinn Vivian, and several geologists evaluated the Sun Dagger site for signs of working of the rock slabs and the cliff surface.

Our later evidence of astronomical alignments in the Chaco buildings inspired scores of other individuals to contribute in-depth knowledge and expertise to the study of Chaco astronomy. Archaeologists, astronomers, archaeoastronomers, physicists, geologists, anthropologists, photographers, art historians, geodesists, laser-scanning technologists, computer graphics modelers, and architects participated in the research. The research teams included volunteers affiliated with the National Science Foundation, the National Geodetic Survey of the National Oceanic and Atmospheric Administration, the U.S. Naval Observatory, Sandia National Laboratory, other scientific agencies and numerous universities. In addition, several members of the National Park Service staff, most notably the archaeology staff, supported and assisted this work.

Among our earliest findings, between 1978 and 1979, we confirmed that the people of Chaco marked the summer and winter solstices, and the equinoxes, by forming vertical light patterns on two spiral petroglyphs at the Sun Dagger site on Fajada Butte (Chapter 1: "A Unique Solar Marking Construct"). In addition to the summer solstice event of the light dagger bisecting the large spiral, at equinox a needle of light bisects the small spiral on the cliff face to the left of the large spiral; and at winter solstice two vertical light shafts fall on the outer edges of the large spiral, bracketing it.

The following year, we found that the astronomers of Chaco also recorded the extremes of the 18.6 year lunar cycle at the Sun Dagger site in patterns of shadow on the larger of the two spiral petroglyphs (Chapter 2: "Lunar Markings on Fajada Butte, Chaco Canyon, New Mexico"). These extremes are known as the major and minor lunar standstills.[4] In commemoration of the major standstill position of the moon, when the moon rises the furthest north in its 18.6 year cycle, a diagonal shadow falls on the left edge of the spiral. Nine and a half years later, when the moon rises at the other extreme of its cycle, the minor standstill position, its shadow falls on the center of the spiral. In each of these extremes of the moon's cycle, its shadow aligns with a pecked groove (Chapter 3: "Astronomical Markings at Three Sites on Fajada Butte"). The nine to ten year passage of time between the lunar standstills correlates with the nine and a half turns of the spiral. Moreover, as the moon progresses from minor to major standtstill over this time, its shadows cross the left side of the spiral with its nine and a half turns.

Between 1982 and 1984, we found that the Chaco people created seven other solar markings on distinctive petroglyphs on Fajada Butte. One of these markings precisely records solar noon each day. The six other markings record solar noon with unique patterns at the summer and winter solstices and equinox (Chapter 3: "Astronomical Markings at Three Sites on Fajada Butte").

Between 1984 and 1997, we conducted surveys of the architecture of Chaco with the help of the National Geodetic Survey. These surveys revealed that the people of Chaco expressed their interest in astronomy not only in solar and lunar light markings but also in the development of an extensive solar-lunar pattern of their primary architecture (Chapter 5: "The Primary Architecture of the Chacoan Culture: A Cosmological Expression"). Twelve of the fourteen major buildings, located in and near Chaco Canyon, are oriented to the extremes and mid-positions of the cycles of the sun and the moon. In addition, the bearings between many of the fourteen buildings are also on alignments to these positions of the sun or moon. This regional pattern of solar and lunar alignments is symmetrically formed about the cardinally organized central complex of Chaco Canyon. It extends approximately thirty miles east-west and forty miles north-south. Our surveys further uncovered a solar-lunar internal geometry of the fourteen major buildings (Chapter 5: "The Primary Architecture of the Chacoan Culture" and Chapter 6: "Chacoan Architecture: A Solar-Lunar Geometry").

In the mid-1980s, our study of the Chaco roads revealed that the thirty-five mile Great North Road was built as a cosmographic expression. This engineered road, generally thirty feet wide and in places expressed in multiple parallel routes, extends north from Chaco Canyon to the badlands of Kutz Canyon. Here it drops down the canyon's steepest edge. Finding no evidence of residential structures on the road or of other functional use, we concluded that it was built to commemorate the direction north and a dramatic topographic feature of the north (Chapter 7: "The Great North Road: A Cosmographic Expression of the Chaco Culture of New Mexico").

In 1989 we discovered physical deterioration of the Sun Dagger site, due to the intense visitation it received since our first recording of it in 1978. We found that a recent pivoting movement of the middle of the three slabs of approximately one and half inches had significantly changed the summer solstice Sun Dagger (Chapter 8: "Changes in

Solstice Marking at the Three-Slab Site, New Mexico, U.S.A."). Our later documentation in 1989 would show that this disturbance also significantly changed the equinox and winter solstice markings. (We would learn in 2005 that further changes at the site affected the position of the eastern slab and thereby significantly shifted the lunar markings.) Seeing this deterioration of the site greatly accelerated the Project's twenty-five-year effort to record it, digitally and accurately, in a three-dimensional interactive computer model. Advances in technology and remarkable expertise finally matched the challenge, helping the Project to produce a successful model in 2006. This model precisely replicates the Sun Dagger site as we first recorded it. In addition, the model's interactive tools allow researchers to conduct extensive studies of the site's astronomical functioning and to explore possible scenarios of its original development (Chapter 9: "The Sun Dagger Interactive Computer Graphics Model: A Digital Restoration of a Chacoan Calendrical Site").

Based on our research findings, we suggest that concepts of cosmology structured and directed the works of the people of Chaco. Over two and a half centuries, they devoted extraordinary efforts to designing and building a physical world carefully aligned to the cycles of the sun and the moon. In an expression of their cosmological concepts they created in Chaco's harsh desert a center of sacred architecture and a vast complex of its astronomical relationships.

This is a relatively new understanding of Chaco. We arrived at it through thirty years of documentation and analysis of extensive data, which are fully presented in these papers. The volume invites readers to conduct their own analysis of our methodology and material, and to reflect on our conclusions, and to perhaps draw their own.

UNDERSTANDING CHACO AS A CENTER OF ASTRONOMY AND COSMOLOGY

Seeing Chaco as a center of astronomy and cosmology did not come to us in isolation from the work of other scholars or from the words of Pueblo historians and educators. First, the field of archaeoastronomy, most especially in the last 40 years, has revealed that numerous ancient sites, as well as the practices of traditional societies, are devoted to the commemoration of celestial cycles. These traditional practices draw the order of the cosmos onto the earth's surface in the alignments of monumental works, and in such smaller expressions as sand paintings, petroglyphs and carved stelae. The works of two brilliant pioneers in these studies had become available as I began my studies. Gerald Hawkins' early insights into the alignments of Stonehenge initiated much of today's understanding of such practices.[5] Anthony Aveni's analysis of Mesoamerican sites greatly deepened the perception of cosmology expressed in their alignments.[6]

Archaeologist Alonso Mendez and his colleagues have recently shown that the Maya site of Palenque incorporates several aspects of solar-lunar astronomy in its architecture and iconography that significantly parallel our findings at Chaco.[7]

Several scholars, independent of the Solstice Project, have contributed critical insights to Chaco as a center of cosmology. Of special interest are the highly significant findings by astronomer John McKim Malville of solar and lunar alignments and symbolic expressions.[8] For instance, his study of Chimney Rock, an outlying Chaco building set high on a precipice in southwestern Colorado, revealed another alignment by the Chaco people to the moon rising at its major standstill.

Readers will also find references to the groundbreaking geometric and cosmographic studies by John Stein, Richard Friedman, and John Fritz; to the astute archaeological insights of Michael Marshall, Stephen Lekson, and Thomas Windes; and to many others still at work on aspects of this ancient cosmology.

As we grew in understanding Chaco as a center of extensive astronomical expression, historians and educators of today's Pueblo culture conveyed that Chaco is a sacred center in their history and that its sites and constructions have spiritual significance. In the documentary *The Mystery of Chaco Canyon*, produced by the Solstice Project, several Pueblo individuals said that the people of Chaco had great knowledge and power, expressed often in power and control over natural life.[9] They also suggested that this power extended to "power over people" and, further, that the sites may have been deliberately sealed and closed for reasons associated with abuse of that power.

LEARNING ABOUT CHACO FROM THE PUEBLO PERSPECTIVE

At critical stages of our work, several Pueblo individuals provided insights into their cosmology that paralleled our findings. One of these individuals was the late Dr. Alfonso Ortiz, an anthropologist and member of Ohkay Owingeh (formerly San Juan Pueblo). In 1978 he saw our photographs of the solar markings at the Sun Dagger site, as we had just begun to find evidence of lunar markings at the site. He immediately, without knowing this, said that "where the sun is so marked so would be the moon," noting the complementary roles of the sun and moon in Pueblo cosmology.[10] Dr. Ortiz also observed that the Sun Dagger site "would have been one of the central concerns of the Chacoans' lives and there would be people there on a regular basis praying, meditating, leaving offerings, and making observations."

A few years later, we shared our first documentation of the Great North Road with Dr. Ortiz. We were finding no utilitarian purpose for the road, but rather that it appeared to be constructed as a commemoration of the direction north and an important topographic feature, the steepest edge of a badlands canyon. Dr. Ortiz said the word in Tewa for road translates as "channel for the life's breath." This thought gave expression to our growing perception that this and certain other Chaco roads, though extremely elaborate constructions, were built as symbolic representations. Several years later, based on Pueblo cosmology today, anthropologist Fred Eggan,[11] archaeologist Edmund Ladd (Zuñi) and historian Paul Pino (Laguna Pueblo) suggested that the Chacoans' roads may have been expressions of their spiritual worldview.[12])

Early in our studies, an unusual petroglyph near the Sun Dagger site suggested symbolic significance in Chaco architecture. It depicts the design of Pueblo Bonito and possibly its relationship to the seasonal cycle of the sun, in the three turns of a spiral etched above the building's design (Chapter 4: "Pueblo Bonito Petroglyph on Fajada Butte: Solar Aspects"). Pueblo Bonito's outline is in the form of a bow crossed by an arrow that is aligned with its north-south midwall. This arrow appears to point above the building and to the south, to the sun at noon.

Dalton Taylor of Hopi responded to photographs of this petroglyph saying "it was more important than the building itself." These words bespoke what we would soon find in our survey of Chaco's primary architecture: the symbolic value of these large buildings is perhaps more significant than their physical mass.

Commenting on the absence of evidence of a written record or a plan expressing the Chaco people's cosmology, David Warren, ethnohistorian and member of the Santa Clara Pueblo, estimated that "in the oral tradition" and "in the spiritual ceremonial systems, there was the inherent instruction on how to lay out (the buildings and roads) according to cosmological requirements."[13]

CHANGING VIEWS OF THE MEANING AND PURPOSE OF CHACO

Over the decades when we were at work in Chaco, several archaeologists of the National Park Service Chaco Research Center at the University of New Mexico arrived through their studies at conclusions that complement the essence of our findings. Rather than seeing Chaco, as they had originally hypothesized, as primarily a trade and redistribution center, their analyses led to a perception of it as a center for religious pilgrimage.[14,15] Studies since have shown a low resident population in Chaco (estimating approximately no more than 2000 people at any one time) and more evidence of extensive ceremonial breakage of pots, further affirming this new understanding of Chaco.[16] The near total absence of hearths and household trash associated with the large buildings suggest that they held few or perhaps no residents.[17]

John Stein and Stephen Lekson observed that the Chacoans' buildings were an expression of their "concepts of the cosmos" and that Chacoan architecture is a "common ideational bond" across "a broad geographic space."[18] Corresponding to this perception, British archaeologist Colin Renfrew described Chaco as "a location of high devotional expression."[19] He suggested that gatherings at such a center might take place "on occasions of calendrical significance." And in addition he noted that notes such a place might contain "features, axes, and orientations of cosmological significance."

Archaeologist James Judge, formerly director of the Chaco Research Center, and astronomer J. McKim Malville wrote that the social organization of Chaco, including the coordinating of massive ritual pilgrimages to the canyon, may have been determined by a solar-lunar calendar.[20] They pointed to the prediction of lunar eclipses (an expertise that would most likely include knowledge of the lunar standstill cycle) as a basis of power and control by a Chaco priest leadership.

The work of John Stein, Dabney Ford and Richard Friedman has shown that a rigorous geometric progression occurred over 200 years in the development of Pueblo Bonito, the Chaco building that held the largest concentration of ritual objects.[21] They suggest that cosmological planning by the builders caused near constant construction and reconstruction of its major elements over two centuries, including extensive opening and closing of its more than thirty kivas. Stein has also shown that in the later phases of Chaco construction, the buildings are designed with little or no open areas or plazas, and that what openings and doorways they had were eventually sealed.

In recent years, we have come to realize that Chaco's place in the world must have been dynamically related to the great region of Mesoamerica to the south. It has long been known through evidence of parrot feathers and copper bells found in the Chaco buildings, that the people there had contact with certain cultures of the south. The extent of this contact and its ideological content are unknown. Now, however, we can see that the cosmological expression evident in Chaco's monumental architecture resonates with the large ceremonial centers of Mesoamerica. At these centers, celestial events appear to have influenced rulership and iconographic content, as well as determined alignments and geometries of large temples and pyramids.

Our recent discussions with scholars of Mesoamerica, Johanna Broda, Flora Clancy, Christopher Powell, and Alonso Mendez, are pointing to the possibility that the Chaco world was developed as an expression of the ancient Americans' knowledge and concern with the integration of latitude, astronomy and geometry. From different perspectives, scholars have increasingly shown that numerous sites of Mesomerica, as well as the 260 day calendar, are related to the ancient Americans' concern with latitude.[22] Most notable is the site of Alta Vista, on the Tropic of Cancer.[23] Here, Aveni *et al.* have shown that this site, an outpost of the Teotihuacan culture of Central Mexico, was constructed purposefully at the Tropic of Cancer, where the passage of the summer solstice sun and of the zenith sun converge.

I have noted that the solar-lunar geometries of the major Chaco buildings correspond with special geometries and that this correspondence occurs uniquely at Chaco's latitude.[24,25] I suggest that this integration of astronomy and geometry, distinctive to Chaco's latitude, may have been a motivating factor for ancient peoples to develop such a primary center of astronomy and cosmology in a remote and quite inhospitable canyon. Perhaps the canyon's topography defined a center place, but in addition, its particular latitude provided opportunities to unite time and space, sun and moon, with aesthetic geometries.

These thoughts have opened the way to further questions about Chaco's role within the relationship between north and south in ancient America: Did Chaco represent to the cultures of the south, which enjoyed lush tropical lands, a place of the north, where spareness and harsh extremes created an opposite environment? And did the Chaco people appreciate the complementary qualities of fertility and abundance in the lands to the south? In Chaco's clear atmosphere, the moon and sun would dramatically traverse the clear skies. These qualities may have been known throughout a great geographical region. Perhaps the barren land of the San Juan Basin held symbolism to that wider area of a netherworld, or at the very least, offered a space of austerity and reverence.

A place so devoid of material support, with its uniquely demanding environs, seems to have invited the construction of enormous buildings without residents, and roads without transport and trade. Rather than supporting residential life, Chaco's major constructions required hard labor from thousands of people in severe conditions. This devotion seems to have been motivated by a cosmological vision. The depths of the vision are probably far beyond our understanding. And yet we know now that the Chaco structures appear to have been built to engage and hold the cycles of the sun and moon. So we can sense in some small manner the compelling purpose of the Chaco people to undertake, over twelve generations, an astonishing physical effort to make manifest their vision.

ACKNOWLEDGMENTS

The Solstice Project's research papers and several of its abstracts originally appeared in journals such as *Science and Archaeoastronomy,* and in volumes published by the University of New Mexico Press, University of Utah Press and Cambridge University Press. Most were presented to conferences such as the 1981, 1986, 1990, and 1993 Oxford International Conferences on Archaeoastronomy, the American Astronomical Society annual meetings, and most recently the 2006 Biennial Southwest Symposium on "Acts of History: Ritual, Landscape, and Historical Archaeology."

Philip Tuwaletstiwa, a geodesist and member of the Hopi tribe who has offered valuable insights into the perceptions and concerns of the Pueblo people about Chaco as a sacred center, contributes the foreword to these papers. Since 1983, Tuwaletstiwa has assisted the Solstice Project's research. In the early 1980s as deputy director of the National Geodetic Survey, he provided extensive technical assistance to the Project's survey of the Chaco major buildings. Over twenty-five years, Tuwaletstiwa has also provided important cultural and scientific insights and guidance to the Project.

We are deeply indebted to scores of individuals who devoted generous hours of probing analysis and dogged efforts of data reduction—the hard work behind the findings presented here. A few individuals were especially indomitable in their devotion to this research: Rolf Sinclair, Volker Zinser, LeRoy Doggett, Michael Marshall. Perhaps most helpful was their persistent and thoughtful skepticism as we opened our minds to a new paradigm of an ancient American culture. I also feel special gratitude to several people who early in this work were eloquent and inspiring to me in their understanding that Chaco reveals so much more than evidence of material adaptation—that the Chaco works are expressions of cosmology; these are Alfonso Ortiz, Fred Eggan, Edmund Ladd, Harold Littlebird, Michael Marshall, Paul Pino, David Warren, and John Stein.

Alan Price, James Holmlund and Joseph Nicoli, and the dedicated work of several others, recently made possible the remarkable computer reconstruction of the Sun Dagger site. William Stone of the National Geodetic Survey continues his rigorous geodetic surveys in assistance to the Project's study of Chaco. Dabney Ford, Chief of Cultural Resources of Chaco Culture National Historical Park, gave especially thoughtful assistance to the work of the Solstice Project in the years 1985 to 2007 for our many projects of surveying the Chaco buildings and the Sun Dagger site. Since 1978 the wisdom and generosity of William and Mary Lou Byler gave heartfelt energy to the Project's three decades of work.

In producing this collection of papers, we are grateful for the encouragement and thoughtful guidance of many individuals, among them: Carlotta Bird, Flora Clancy, James Faris, Howard Fisher, Craig Johnson, Mary Judge, Mac MacLaren, Bobbe Needham, Michael Pertschuk, Richard Polese, Rina Swentzell, Phillip and Judith Tuwaletstiwa.

NOTES

1. Aveni, Anthony, *Skywatchers of Ancient Mexico,* University of Texas Press, 1980.

2. Anthropologist Ken Hedges showed his slides at the 1977 Annual Meeting of the American Rock Art Research Association in Tempe, AZ. Hedges, Ken, "Rock Art in the Pinon Forests of Northern Baja California," *AIRA,* Vol. 3, ed. A.J. Bock, F. Bock, and J. Cawley, 1977, pp. 1-8.

3. Williamson, R.A., H.J. Fisher, A.F. Williamson, and C. Cochran, "The Astronomical Record in Chaco Canyon, New Mexico," in *Archaeoastronomy in Pre-Columbian America*, ed. A.F. Aveni, University of Texas Press, 1975, pp. 33-43.

4. The moon's standstill cycle is longer (18.6 years) and more complex than the sun's cycle, but its rhythms and patterns also can be observed in its shifting positions on the horizon, as well as in its relationship to the sun. In its excursions each month it shifts from rising roughly in the northeast to rising roughly in the southeast and from setting roughly in the northwest to setting roughly in the southwest, but a closer look reveals that the envelope of these excursions expands and contracts through the 18.6-year standstill cycle. In the year of the major standstill, this envelope is at its maximum width, and at the latitude of Chaco, the moon rises and sets approximately 6.1° north and south of the positions of the rising and setting solstice suns. These positions are the farthest to the northeast and northwest and southeast and southwest that the moon ever reaches. In the year of the minor standstill, nine to ten years later, the envelope is at its minimum width, and the moon rises and sets approximately 6.7° within the envelope of the rising and setting solstice suns.

5. Hawkins, Gerald, *Stonehenge Decoded*, Doubleday, 1965.

6. Aveni, op. cit.

7. Mendez, Alonso, Edwin L. Barnhart, Christopher Powell, and Carol Karasik, "Astronomical Observations from the Temple of the Sun," *The Journal of Astronomy and Culture*, Vol. XIX, 2005, pp. 44-73. In this work, Mendez *et al.* demonstrate that the Temple of the Sun of Palenque incorporates alignments to the sun at equinox, solstice, and zenith passage, and to the moon at its major standstill. The solar alignments are accompanied by light markings as well. These researchers also note geometries in the solar-lunar relationships of the Temple of the Sun that have the interesting property of the 3-4-5 right-angle triangle. Powell, at two Solstice Project seminars (Santa Fe, September 2005 and Chaco Canyon, May 2007), has proposed that there are yet other geometries of special note in the solar-lunar relationships of Palenque, and he hypothesizes that the Maya selected the site for its particular latitude where these astronomical-geometric relationships could be architecturally commemorated. I have similarly noted that "it may be that Chaco Canyon was selected as the place, within the broader cultural region of Mesoamerica, where the relationships of the sun and the earth, and the sun and the moon, could be expressed in geometric relationships that were considered particularly significant" (Sofaer in Chapter 5: *The Primary Architecture of the Chacoan Culture*, note 18).

8. Malville, J. McKim and Claudia Putnam, *Prehistoric Astronomy in the Southwest*, Johnson Books, 1989.

9. Edmund Ladd, archaeologist and member of Zuñi Pueblo; Paul Pino, historian of Laguna Pueblo; and Petuuche Gilbert, historian of Acoma Pueblo, make these statements in *The Mystery of Chaco Canyon*, a documentary produced by the Solstice Project and narrated by Robert Redford, telecast by PBS, 2000-2007.

10. *The Sun Dagger*, a documentary produced by the Solstice Project and narrated by Robert Redford, telecast by PBS, 1982-1988.

11. Fred Eggan, personal communication 1990.

12. Edmund Ladd and Paul Pino in *The Mystery of Chaco Canyon*.

13. David Warren in *The Mystery of Chaco Canyon*.

14. Judge, W. James, "New Light on Chaco Canyon," in *New Light on Chaco Canyon*, ed. David Grant Noble, School of American Research Press, 1984, pp. 1-12.

15. Judge, W. James and John D. Schelberg, eds., *Recent Research on Chaco Prehistory*, Division of Cultural Research, U.S. Dept. of the Interior National Park Service, 1984.

16. Lekson, Stephen H., ed., *The Archaeology of Chaco Canyon, An Eleventh Century Pueblo Regional Center*, School of American Research Press, 2006.

17. Richard Friedman in *The Mystery of Chaco Canyon*.

18. Stein, John and Stephen H. Lekson, "Anasazi Ritual Landscapes," in *Anasazi Regional Organization and the Chaco System*, David E. Doyle, ed., Maxwell Museum of Anthropology, Anthropological Papers 5, University of New Mexico, 1992

19. Renfrew, Colin, "Chaco Canyon, A View from the Outside," in *In Search of Chaco*, David Grant Noble, ed., SAR Press, 2004, pp. 101-106

20. Judge, W. James and J. McKim Malville, "Calendrical Knowledge and Ritual Power," in *Chimney Rock*, ed. J. McKim Malville, Lexington Books, 2004, pp.151-162.

21. Stein, John R., Dabney Ford, and Richard Friedman, "Reconstructing Pueblo Bonito," in *Pueblo Bonito: Center of the Chacoan World*, Jill E. Neitzel, ed., Smithsonian Institution Press, Washington, D.C., 2003.

22. Broda, Johanna, "Zenith Observations and the Conceptualization of Geographical Latitude in Ancient Mesoamerica: A Historical Interdisciplinary Approach," in *Viewing the Sky Through Past and Present Cultures*, ed. Todd W. Bostwick and Bryan Bates, Pueblo Grande Museum Anthropological Papers, Number 15, 2006.

23. Aveni, op.cit.

24. See note 7.

25. I elaborated on this proposal in presentations to Solstice Project Seminars: "Chaco/Mesoamerica," Grants, NM, 2000, and "Chaco/Mesoamerica II," Santa Fe, NM, 2005.

Light Markings and Petroglyphs on Fajada Butte

1

A Unique Solar Marking Construct

An Archeoastronomical Site in New Mexico Marks the Solstices and Equinoxes

Summary: *An assembly of stone slabs on an isolated butte in New Mexico collimates sunlight onto spiral petroglyphs carved on a cliff face. The light illuminates the spirals in a changing pattern throughout the year and marks the solstices and equinoxes with particular images. The assembly can also be used to observe lunar phenomena. It is unique in archaeoastronomy in utilizing the changing height of the midday sun throughout the year rather than its rising and setting points. The construct appears to be the result of deliberate work of the Anasazi Indians, the builders of the great pueblos in the area.*

NEAR THE TOP OF AN ISOLATED BUTTE in Chaco Canyon, New Mexico, three large stone slabs collimate sunlight in vertical patterns of light on two spiral petroglyphs carved on the cliff behind them. The light illuminates the spirals each day near noon in a changing pattern throughout the year and marks the solstices and equinoxes with particular images. At summer solstice a narrow vertical form of light moves downward near noon through the center of the larger spiral. At equinox and winter solstice corresponding forms of light mark the spirals. We found that the relationship between the shape and orientation of the slabs and the resultant light patterns on the cliff is a complex one and required a sophisticated appreciation of astronomy and geometry for its realization.

The site is unique in employing the varying height of the midday sun during the year to provide readings of solar declination. In this respect it is clearly different in concept from the many archeoastronomical sites throughout the ancient New and Old Worlds that tell the passage of the year by marking the rising and setting points of the sun and moon.[1]

The Anasazi Indians occupied Chaco Canyon from about A.D. 400 to 1300 (Figure 1). In this arid and unproductive region, these early inhabitants left evidence of a skilled and highly organized society. They constructed multistory pueblos and large ceremonial centers, and developed extensive systems of roads, irrigation, communication, and trade.[2]

Anna Sofaer
Volker Zinser
and
Rolf M. Sinclair

Originally published in **Science,**
*Volume 206, Number 4416
19 October 1979*

(Volker Zinser is an architect in Washington, D.C. Rolf Sinclair served in the National Science Foundation Physics Program.)

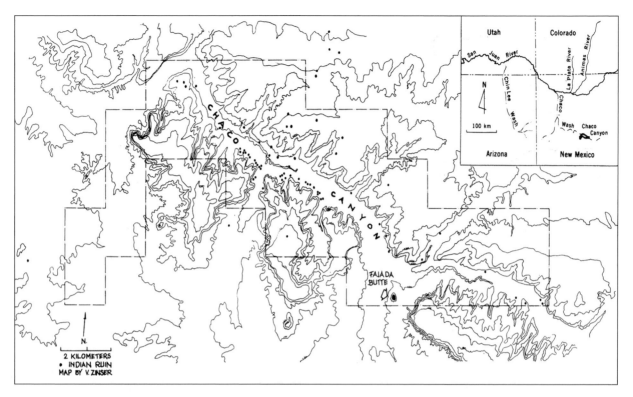

FIGURE 1. *Map of Chaco Canyon National Monument showing the location of the Indian ruins (circles) and Fajada Butte, and (inset) the Four Corners area of the Southwest United States.*

The Anasazi had established an accurate calendar for agricultural and ceremonial purposes. To do this they determined the recurrence of the solstices and equinoxes.[3] This astronomical knowledge was commemorated in the design and alignment of major buildings.[4, 5] The precision of the historic Pueblo calendar has been described in a recent study of its synchronization of the monthly lunar cycles with the annual solar cycle.[6]

Fajada Butte stands prominently in the south entrance of Chaco Canyon (Figure 2), rising 135 meters above the valley floor to an elevation of 2018 meters. The butte is difficult to climb, and there is neither water nor soil on it. Yet from bottom to top are many examples of Indian rock art carved and painted on the cliffs,[7] the ruins of a number of small Pueblo buildings, and countless pottery shards. This concentration of remains attests to its active use by the Indians. The butte is a natural site for astronomical observations, with its clear views to distant horizons.[4]

The last 10 m to the summit of the butte on the southeast face is formed by a vertical cliff with a narrow ledge at its foot. The assembly we will describe consists of an unusual arrangement of three stone slabs 2 to 3 m in height, standing on edge on this ledge and leaning against the cliff. The slabs are slightly off the vertical (Figure 3) and fan out radially (Figure 4). They are close together, separated by only 10 centimeters at their inner edges, but do not touch. The slabs keep the cliff face behind them in shadow except near midday, when the sun shines between them to cast patterns of light.[8]

Two spiral petroglyphs are carved ("pecked") on the cliff immediately behind the slabs (Figure 5). The larger and more prominent spiral is located behind the opening of slabs one and two. It has 9½ turns and is elliptical in shape (34 by 41 cm). The smaller spiral (9 by 13 cm) is above and to the left behind the opening between slabs two and three. It has 2½ turns and a loop extends out of its upper right side. The spirals can be seen in their entirety

24

FIGURE 2. *Fajada Butte from the north. The site described in this article is just out of sight on the southeast summit.*

only from the right of slab one. From this position, the larger spiral appears circular, suggesting that this was the intended viewpoint (Figure 6). The length of time the sun shines between the slabs onto the spirals varies from 18 minutes near noon at summer solstice (when the sun is highest in the sky) to three hours or more at winter solstice (when the sun is lowest).

Not surprisingly, moonlight generally creates the same patterns on the spirals as the sun, on nights when the moon is between first and third quarter. The periodic changes in these patterns reflect the complexity of the moon's apparent motion,[9] and certain combinations of patterns are associated with specific lunar eclipses.

The first observations at the site were made on 29 June 1977 by one of us (A.S.), who then initiated the present studies and was soon joined by the other two authors. In what follows, we will describe in detail our record of the patterns of light on the spirals during an annual cycle of the sun.

We will show how the combination of the patterns and the spirals can be used to determine accurately the time of recurrence of the solstices and equinoxes. We will describe how geologic, archeologic, and geometric studies of the site have led us to conclude that the assembly was constructed by the ancient Pueblos. Finally, we will speculate on how the assembly could have been used to study some aspects of the complex cycles of the moon's motion. [Preliminary reports on this work have been presented at two recent conferences.[10]]

OBSERVATIONS

We made records of the patterns of light on the spirals at monthly intervals from 21 June 1978 (summer solstice) to 21 December 1978 (winter solstice) and at some intervening dates. Photographs were taken from several fixed camera locations at 30-second intervals each day as long as light shone between the slabs onto the cliff. Additional photo-

FIGURE 3. *View of the site from the south. The slabs are numbered as referred to in the text, and their original placement on the cliff as drawn by Blair[13] is indicated.*

graphs were made by moonlight on a few nights. We measured the details of the faces and edges of the rock slabs and determined just what points were casting shadows. From an analysis of this information and theodolite recordings, we made geometric reconstructions of the phenomena.

Solar observations. Throughout the solar cycle two consistent characteristics of the movement of the light patterns make possible this solar calendar:

1) Each day the light is a vertical form and moves primarily in a downward vertical motion, while the sun moves essentially horizontally.

2) As the sun's midday passage lowers in altitude from summer solstice to winter solstice, that downward vertical movement gradually takes place farther to the right on the cliff face.

Summer solstice (21 and 22 June). On this day sunlight first shines through the right opening (between slabs one and two) at 11:05:15 a.m. local apparent time.[11] The light first appears as a spot on the top edge of the outside turn of the larger spiral (Figure 7a). The spot quickly stretches into a thin vertical form, sharply pointed at the tip (Figure 7b). It lengthens downward and descends through the center of the spiral. At the time halfway between the first and the last appearance of light on the cliff face, the form is centered both horizontally and vertically on the spiral (Figure 7c). Thereafter, the pattern continues to move downward (Figure 7d), becomes shorter (Figure 7e), and finally disappears near the bottom of the spiral 18 minutes after its initial appearance.

On the days after summer solstice the light pattern follows the same sequence but is displaced increasingly to the right of the center of the spiral. Recalling that the phenomenon is symmetrical about summer solstice, the sequence of events from May through July is this: the form of light as it descends vertically each day passes at first to the right of center. On succeeding days, it moves slightly left until at solstice it reaches an extremum, passing through the center of the spiral. Thereafter it moves back to the right each day. On 21 July or 22 May, it passes 3.2 cm to the right of center. A shift of 2 millimeters to the right can be detected by comparing photographs taken on solstice day and 25 June, and thus the time of summer solstice is marked within four days, when the sun's declination has shifted only 2 minutes and 3 seconds of arc.

In a related effect at summer solstice, sunlight reaches the cliff face through the left opening (between slabs two and three) for only two minutes to cast a small spot of light, which is not noticeable to the casual observer. Only a few days earlier or later, this spot becomes a clearly visible form and its time is longer by one minute or more. The length of time that sunlight reaches the cliff face through the left opening decreases steadily as summer solstice is approached, from more than two hours in winter to two minutes on 21 June (Figure 8). If we extrapolate linearly to the maxi-

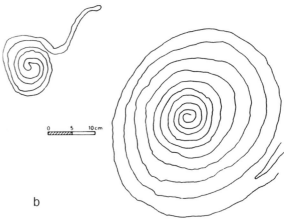

FIGURE 4 (above). (a) Close-up of the slabs from the southwest. (b) Top view of the slabs.
FIGURE 5 (right). (a) Close-up of spirals (artificially highlighted). (b) Tracing of spirals.

mum declination of the sun in A.D. 1000 [23°34.1'⁹], we find that this length of time would then have been only one minute. This minimum is not in itself surprising, since the length of time must vary symmetrically about solstice. But having the minimum time so close to zero is curious and suggests that 1000 years ago it was perhaps even less. We have located those parts of the edges of the slabs that determine how long light shines through the left opening at solstice. A change in the position and height of the top surface of slab two by only 1 to 2 mm would be enough to reduce the time by one minute. Since this change is a reasonable estimate of the effects of weathering and settling over a millennium,[12] it is likely that the time during which light entered this opening was close to zero (or reached zero) at summer solstice and thus provided a second precise marking of summer solstice.

FIGURE 7. *Light pattern moving downward across the larger spiral near summer solstice (23 June 1978): (a) 11:07:00 a.m., (b) 11:10:45 a.m., (c) 11:13:00 a.m., (d) 11:16:45 a.m., and (e) 11:20:45 a.m.*

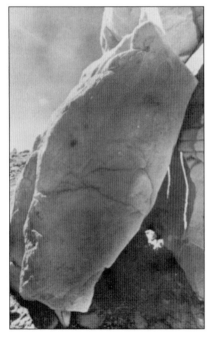

FIGURE 6. *Light patterns on the spirals from the viewing point to the right of slab one on a typical day (28 October 1978, 10:35 a.m.).*

Autumn equinox (21 and 22 September). After summer solstice, both the shape of the light patterns and the paths they follow change each day. The time that sunlight shines through the left opening increases, and the pattern of light grows gradually from a spot to a slender inverted triangle. This form starts each day closer to the smaller spiral and eventually moves down across it, first passing to the left of the center. Each day the downward path moves a little to the right, until at equinox the light form passes through the center of this spiral (Figure 9b), just as the other light form passed through the center of the larger spiral three months earlier at summer solstice (Figure 9a). After equinox, the path continues to move to the right. A week before or after equinox, the form passes clearly to the left or right of the center of the small spiral.

The pattern formed by sunlight shining through the right opening changes gradually from the downward-moving thin vertical form observed at summer solstice (Figure 7 and Figure 9a) to a band of light stretching across the larger spiral (Figure 9b). At equinox this band, as it first spans the spiral, is positioned approximately 7 cm to the right of center, between the fourth and fifth

turns—that is, in the midspace of the nine turns. The daily track of its motion is displaced gradually to the right on successive days.

Winter solstice (21 and 22 December). Between autumn equinox and winter solstice, the pattern produced by sunlight through the left opening still starts each day as a narrow pointed shape and descends downward. During the next hour or two it becomes a vertical band and moves slowly rightward across a portion of the larger spiral. The track of this light form's motion shifts—like the light formed by the right opening—gradually rightward day by day.

At winter solstice markings occur on both petroglyphs. The first involves the left pattern only. Shortly after its first appearance on 21 December, while still a slender pointed shape, it descends through the bottom of the loop extending to the right from the smaller spiral (in a position 7 cm to the right of the equinox light form) (Figure 5b). Then later on the same day both bands of light move so as to "frame" the larger spiral (Figure 9c). This occurs when the left band reaches the left side of the spiral and is in the outer left groove just as the right band crosses the spiral and is in the outer right groove. This framing of the spiral occurs only within one week of solstice.

After winter solstice, the daily change of the patterns of light on the spirals repeats itself in reverse order. By spring equinox (21 March) the

patterns and their motion are the same as at autumn equinox, and by summer solstice the solar cycle is completed.

An overview of the light patterns through the solar cycle reveals relationships between the markings of each season (Figure 9, a, b, and c). The markings of the winter and summer solstices each involve a significant change in the light forms that mark the other quarters. For example, at summer solstice the observer's attention is focused without distraction on the dramatic occurrence of the bisecting light on the larger spiral because no light is formed by the left opening. This phenomenon is

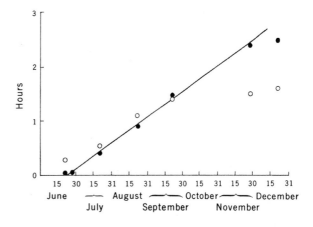

FIGURE 8. *Variation with date of the length of time that sunlight shines on the cliff face between (○) slabs one and two and (●) slabs two and three.*

29

especially striking since light shaped by the left opening develops soon after summer solstice into a distinct vertical form, which eventually bisects the smaller spiral at equinox. In another example of the seasonal interrelationships of the light forms, the light shaped by the right opening, which acts significantly at summer solstice, is joined by the light shaped by the left opening, which acts significantly at equinox, to become the bracketing partners that frame the larger spiral, which is "empty" of light at winter solstice.

Lunar observations. The patterns formed at night by moonlight shining between the slabs are just as clear and as noticeable as those formed by the sun, and we easily recorded them on several nights near full moons (Figure 10). It is not necessary to make a detailed record of these patterns for most positions of the moon, since at a particular point in the sky it will form the same light patterns on the spirals as the sun would at the same point. When the moon's declination is between +23.5° and –23.5° (the solar extremes) we can thus predict the patterns formed by its light by knowing those formed by the sun at the same declination. But the moon's declination can vary outside the solar limits, up to ±28.5° over part of an 18.6-year cycle, and whenever it lies beyond the extremes of the sun's declination we have no corresponding solar data. Since this periodic extreme in the moon's declination will not be reached again until 1987, we cannot yet make direct observations of the patterns formed by the moon at such declinations. As discussed later in this article, extrapolations from the solar data suggest that a significant marking of the maximum lunar declination may occur.

THE STONE SLABS

The three slabs stand on the sloping ledge at the foot of the cliff, each contacting the cliff over only a small area (Figure 11a). On the left of slab three is a supporting buttress of smaller rocks (Figure 4a) and under the right edge of slab one is a small supporting rock. All the slabs and rocks of the

FIGURE 9. *Light pattern on the spirals at the quarter-year points: (a) near summer solstice (26 June 1978, 11:13:15 a.m.), (b) autumn equinox (21 September 1978, 10:50 a.m.); the inset shows the bisection of the smaller spiral by the left light formation; and (c) near winter solstice (22 December 1978, 10:19 a.m.).*

assembly consist of the same soft sandstone as the cliff itself. The slabs are roughly rectangular (2 to 3 m high, 0.7 to 1 m wide, and 20 to 50 cm thick) and weigh about 2000 kilograms each. The outer surfaces and tops are rounded and weathered, the inner surfaces smooth and gently curved with sharp edges (Figure 11b). By comparing the matching details on the facing surfaces of the slabs, it has been determined[13] that these slabs once fitted together to form one block. The place where this block was joined to the cliff face was found by noting the strata and bedding planes and the curvature of the cliff face.[13] By such comparison, the original location of each slab was found to within 1 cm, well to the left of the present locations (Figure 3). The long dimensions of the slabs were then horizontal, and slab one was on top.

Several pieces of evidence rule against the

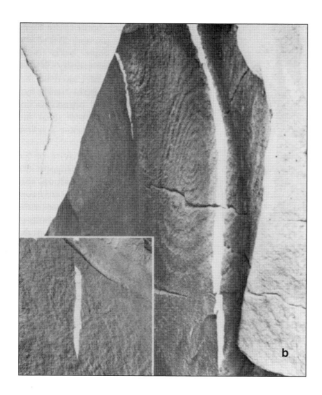

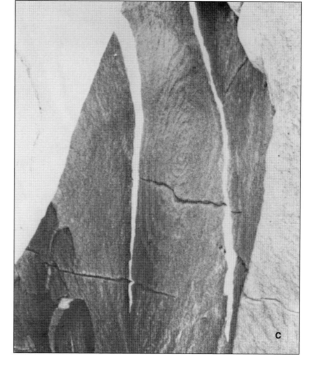

slabs' having fallen into their present positions naturally. First, the slabs would have had to move two meters and more horizontally while the center of gravity of the three together fell only about 80 cm vertically. In particular, the center of slab three by itself is now only 30 cm lower than when it was attached to the cliff. Second, there are no impact marks, either on the cliff face or on the inner edges of the slabs, to suggest a collision. Third, the cliff face above the original location of the slabs shows that another rock mass had broken away from there. This higher rock could not have broken off before the slabs did. Had it broken off with (or after) the slabs, it would have prevented them from falling naturally to their present location. There is no evidence today of this rock mass. Indeed the absence of rubble near the slabs is unusual on the butte, where fallen rock is found below other such cliffs. Fourth, the slabs are set firmly in place on a rocky ledge and are partially supported by buttressing stones. We conclude that moving and setting the slabs in their present position involved deliberate human intervention.

GEOMETRY OF THE ASSEMBLY

The daily patterns of light that highlight special features of the spirals at the quarters of the solar year are formed by several components of the assembly acting as a whole. The placement, size, and shape of the slabs, the orientation of the cliff face, and the positions and sizes of the spirals are all critical. Our calculations and drawings have determined the role and importance of these interlocking factors and show that a change in any one would change or eliminate the images marking the quarters. No one component dominates in such a way that the others could be left to chance.

Consider first the downward motion of the light forms each day and the rightward placement of the light as time progresses from summer solstice. These characteristics of the light patterns' movement are essential to the solar calendar because they provide discernably different markings on a small area of the cliff through the solar cycle. The precise positioning of curved surfaces on the slabs in relation to the cliff face and the sun makes possible the daily vertical movement of light and

31

the seasonal shift in the discretely shaped light forms. The horizontal motion of the sun as it shines against these curved surfaces is transformed into downward vertical movement of the light patterns.

For example, at summer solstice, during the 18 minutes when light shines through the right gap, the sun has moved 4.5° westward almost horizontally across the sky. We would at first glance expect the projected pattern of sunlight to move horizontally across the spiral, shifting to the right on the cliff face. The observed vertical motion is thus in itself surprising. A simple way to picture how this downward motion is effected is as follows: Imagine a long narrow cylinder so oriented that the sun as it moves across the sky can shine through it only briefly onto the cliff to form a spot of light. Immediately below, picture a second cylinder so aligned that sunlight passes through it just as it stops shining through the first one, and now casts a second spot of light immediately below the first. Then continue in a like manner with further cylinders. The result will be that a succession of spots will appear on the cliff, moving downward as the sun moves horizontally. If the cylinders are now joined to form a continuous curved slit, the spot of light on the cliff will move smoothly downward. More generally, a form of light, rather than simply a spot, can be made to move vertically in a similar manner.

Close inspection of the slabs shows that the front left edge of slab one and the right rear edge of slab two form just such a curving slit (Figure 12). As the sun moves across the sky, different portions of the slabs' curved edges come into play to collimate the sunlight. There is a further requirement that these edges be shaped so as to produce the parallel sides and constant width of the downward-moving light form (Figure 7). Further, the slabs and their curves must be located so that the pattern bisects the spiral only at solstice. As the sun lowers after 21 June, the curved surfaces cause the entering light to move rightward on the spiral. A simple vertical opening would not provide the visibly rightward displacement of light in the four days following summer solstice, when the sun has lowered only 2 minutes and 3 seconds of arc. Similar curved surfaces of the left front edge of slab two and the right rear edge of slab three produce the downward motion of the other light pattern at equinox and the rightward shift of this light pattern as time progresses toward winter solstice. In addition to these requirements on the edges of slab two, this slab must be of such a size and so inclined as to shadow the left opening and form the pronounced minimum at summer solstice.

Another feature of the light patterns deserves special notice. The distance from the slabs to the cliff is such that if a pattern of light cast by a small opening between the slabs falls high on the cliff in the winter when the sun is low in the sky, we would expect it to fall on the ground between the slabs and the cliff in the summer. We find, however, that throughout the year, the patterns of sunlight are projected onto only that part of the cliff face around the spirals. This "display area" is just that

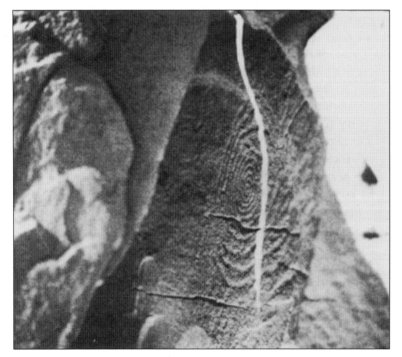

FIGURE 10. *Patterns cast by moonlight (24 June 1978, 2:45 a.m.).*

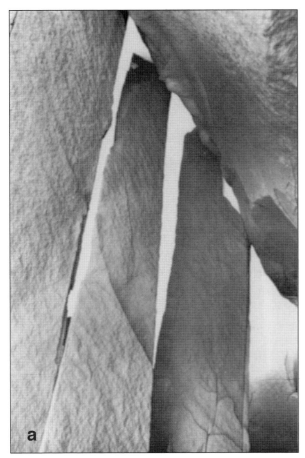 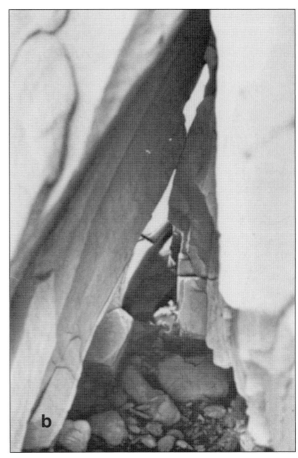

FIGURE 11. *(a) Inside view of slabs looking upward to show area of contact with the cliff. (b) Inside (concave) edges of the slabs.*

viewed comfortably from the right of slab one (Figure 6), which makes it easy to compare patterns at different times of day and of the year. By doing this, the assembly acts like modern instruments that confine the images of interest to a projection screen.

An illustration of how the construct provides information about the solar cycle on a small exhibit space is the contrasting imagery of markings on the larger spiral between the solstices. At winter solstice, when the light forms shaped by both openings bracket the larger spiral, they are equally distant from the center, which was bisected by the light from the right opening six months earlier. That this relationship of both light forms to the center of the spiral occurs only at winter solstice is an indication of the intricately controlled relationship of the slabs.

An important question is whether the slabs or the cliff face show evidence of deliberate shaping. Three areas have been noted by specialists as possibly showing signs of modification. Parts of the top edges of slabs one and two have a surface texture different from that of the rest of the slabs and could have been shaped by pecking.[14, 15] We find that these surfaces are the portions of the slabs that determine the right sides of the downward-moving light forms between summer solstice and equinox, and one of these surfaces is a critical factor causing the pronounced minimum at summer solstice. Another surface that may have been worked is that on which the spirals are pecked. This portion of the cliff face is different from the adjoining areas in shape and surface texture and may have been dished out and smoothed.[16]

33

FIGURE 12. *Formation of the descending light pattern at summer solstice by slabs one and two.*

CULTURAL BACKGROUND

Several factors show that the Anasazi inhabitants of Chaco developed the construct between A.D. 900 and 1300 (the approximate date of Pueblo abandonment of the canyon) and indicate that the specific time was between A.D. 950 and 1150, the period of greatest population and development in the canyon.[17] The sun-watching practices of the primarily agrarian settlements of Pueblo culture have been extensively documented in ethnographic reports as the means of setting planting and ceremonial calendars. These practices were frequently performed at shrines located on the top or near the top of mountains and buttes within the vicinity of Pueblo settlements.[18, 19]

Certain rock art sites of the ancient Pueblos are reported to mark solar positions by the placement of designs to receive shadow and light formation at the rising and setting of the sun at solstice or equinox, and one such site includes a spiral design.[20] Two petroglyph sites on Fajada Butte are marked with shadow and light changes at the time of solar noon at summer solstice, and one of these includes a spiral design.[21] The spiral is frequently found in association with ancient Pueblo petroglyphs of sun imagery.[22, 23] It is identified with the Anasazi rock art style prior to A.D. 1300.[23] Examples of architecture of the same period have openings that channel light so that it shines on key features of the structures such as doorways, niches, and corners at the solstices and equinoxes.[4, 24, 25]

Indications that the Fajada Butte solar marking construct was developed within the period A.D. 950 to 1150 are the planning skills and solar interests exhibited by the Chaco occupants at that time. During this phase of extensive trade and ceremonial activity in the canyon, complex systems of roads, communications, irrigation, and ceremonial architecture were developed. The largest structure of the entire Anasazi culture, Pueblo Bonito ("beautiful village"), a five-story 800-unit building, was built with its primary elements of design precisely aligned to the rising and setting of the equinox sun and the daily noon position of the sun.[7] Pottery shards of the Bonitians are prevalent on and below the Butte—and several have been found within the immediate vicinity of the slabs. The occupants of Chaco at this time of thriving development clearly had the planning ability and organizational skill to move the assembly of slabs and set them in their present solar marking alignments.

Indications of specific activity of the ancient Pueblos at the site suggest the need for future archeological study. It is possible that there was a single structure located about 50 feet to the west of the slabs and a basal retaining wall built to prevent erosion a few feet down the slope from the slabs.[17] Further study might indicate with more certainty the specific group of the Anasazi culture who built the solar marking construct and provide further information about their use of the site.

POSSIBLE SIGNIFICANCE OF THE LUNAR MARKINGS

The assembly may have been used for lunar observations. Such use would be consistent with the Pueblo culture's emphasis on the moon, which is seen in a dual role with the sun, and association of danger with the occurrence of lunar eclipses.[26] The Indians of the Southwest knew of the 18.6-year cyclic extremes in lunar declination and incorporated that knowledge in a major archeological site.[27]

We extrapolated some features of the record of solar observations to estimate what lunar light patterns would occur on the large spiral at the extreme maximum declination of the moon. We found that the length of time that light from the sun (and equally well from the moon) shines through the right opening is a linear function of declination near 23.5° (Figure 13) and extrapolates to zero close to +28.5°. It is thus likely that as the moon reaches its maximum declination, the length of time that moonlight shines on the cliff would be close to zero or would reach zero. Interestingly, such a marking of the lunar extreme would be similar to the indication of the sun's extreme declination at summer solstice by the pronounced minimum of the light form shaped by the left opening.

In observing the lunar markings on the spirals over the 18.6-year cycle, one would be particularly interested in those times when the declination of the full moon is equal and opposite to that of the sun. This occurs once every nine to ten years, within two weeks of the solstice. The moon would then cast on the spirals at night the light patterns characteristic of the solstice, opposite the patterns that would be marked within two weeks by the sun during the day. Correspondingly, a full moon would cast the equinox pattern within two weeks of vernal or autumnal equinox every nine to ten years, but out of phase with solstitial eclipses. These striking alternations of patterns would indicate the alignment of sun, earth, and moon that makes possible a midwinter or midsummer solstitial or an equinoctial eclipse. We should, in this connection, mention the interesting parallel

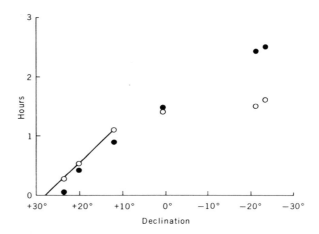

FIGURE 13. *Variation with declination of the length of time sunlight shines on the cliff face between (○) slabs one and two and (●) slabs two and three.*

between the number of turns in the larger spiral (ten on the left and nine on the right) and the alternation of 10 and 9 years in these eclipse cycles, as well as in the moon's transition from minimum to maximum declination. We speculate that the spiral may record this knowledge. Of additional interest is the correspondence between the total number of spiral rings and the 19-year cycle of occurrence of the full moon on the same solar date.

CONCLUSION

The Fajada Butte construct is unique in archaeoastronomy as the only device known to use the passage of the midday sun to create a solar calendar. Precisely planned relationships of curved rock surfaces make possible the transformation of the horizontal movement of the sun into vertical movement of light forms that provide accurate measurements of solar positions. Thus, with no reliance on foresights or horizon markers, the construct is a self-contained instrument that records the sun's changing declination. It shows the times of solstice and equinox in vividly symbolic imagery of light and shadow and provides solar (and lunar) information at other times in the year. While the construct achieves its results with an accuracy comparable to that of the large monuments and structures of other ancient cultures, it does so with a subtle integration with nature that is typical of the North American Indian culture.

Finally, we wish to note that the soft sandstone material of the construct is fragile and can easily be damaged. We hope that efforts will be made to preserve this unusual feature of the Native American heritage.

REFERENCES AND NOTES

1. For a review of archaeoastronomy and of the astronomical terms and concepts used here, see, for example, E. C. Krupp, ed., In *Search of Ancient Astronomies* (Doubleday, New York, 1977).

2. H. M. Wormington, *Prehistoric Indians of the Southwest* (Museum of Natural History, Denver, 1947); G. Vivian and P. Reiter, *The Great Kivas of Chaco Canyon* (Univ. of New Mexico Press, Albuquerque, 1965); R. G. Vivian, "Aspects of prehistoric society in Chaco Canyon, N.M.," thesis, University of Arizona (1970) (available from University Microfilms, Ann Arbor, Mich.); F. Early, *Chaco Canyon, A Study Guide* (Museum of Anthropology, Arapahoe Community College, Littleton, Colo., 1976); F. Folsom, *The New York Times Magazine* (20 August 1978), pp. 18-19, 34-35, 37-38.

3. F. H. Ellis, in *Archaeoastronomy in Pre-Columbian America*, A. F. Aveni, ed. (Univ. of Texas Press, Austin, 1975), pp. 59-88; L. Spier, *Mohave Culture Items* (Northern Arizona Society of Science and Art, Flagstaff, 1955), pp. 16-33; A. M. Stephen, in *Hopi Journal*, E. C. Parsons, ed. (Columbia University Press, New York, 1936; reprinted by AMS Press, New York, 1969).

4. R. A. Williamson, H. J. Fisher, D. O'Flynn, in *Archaeoastronomy in Pre-Columbian America*, A. F. Aveni, ed. (Univ. of Texas Press, Austin, 1975), pp. 33-43.

5. R. A. Williamson, *Smithsonian* 9, 78 (October 1978).

6. S. C. McCluskey, *Journal of the History of Astronomy* 8, 174 (1977).

7. A. Sofaer and J. Crotty, unpublished observations.

8. Light can also shine in from the right of slab one for a short time at sunrise.

9. E. C. Krupp, op. cit., pp. 1-37.

10. A. Sofaer, V. Zinser, R. M. Sinclair, papers presented at the Symposium of the American Rock Art Research Association, The Dalles, Oregon, May 1978, and at the conference on Archaeoastronomy in the Americas, Santa Fe, N.M., June 1979. The work was described in K. Frazier, *Science News* 114, 145 (1978).

11. Local apparent time is used throughout this article. This is the time that would be told by a sundial at a particular location on a given day, with noon occurring when the sun is on the meridian due south at its maximum altitude for the day. This time differs from local mean ("clock") time by at most a few minutes. The maximum daily altitude of the sun varies sinusoidally during the year from 77.5° to 30.5° at Fajada Butte (latitude, 36°N).

12. B. Wachter, private communication.

13. R. Blair, private communication.

14. R. G. Vivian, in a memorandum dated 11 September 1978, stated: "The top edges of the two largest slabs (the right and center slab when facing the cliff) *may* show some evidence of shaping through pecking and grinding. These two surfaces are rather flat and are covered with small dimpled marks suggestive of pecking."

15. C. B. Hunt, in a letter dated November 1978, said that the near edge of the middle slab "looks as if it had been pecked and rubbed."

16. R. G. Vivian, in 11 September 1978 memorandum, stated: "The surface of the cliff face onto which the spiral had been pecked did appear to have been smoothed—possibly through grinding. This surface was somewhat dished or concave, an attribute that may have been achieved through grinding."

17. R. G. Vivian, private communication.

18. F. H. Ellis, private communication.

19. M. C. Stevenson, *Bureau of American Ethnology Annual Report* 23 (1904).

20. R. A. Williamson, paper presented at the conference on Archaeoastronomy in the Americas, Santa Fe, N.M., June 1979. It is interesting to note that parallels in this type of solar marking are found among California Indian cultures (see T. Hudson, G. Lee, K. Hedges, *Journal of California Anthropology*, in press); one example in the report by Hudson *et al.* includes a spiral petroglyph crossed by shadow at winter solstice.

21. A. Sofaer, in preparation.

22. F. H. Ellis, private communication.

23. P. Schaafsma, private communication.

24. J. E. Reyman, *Science* 193, 957 (1976).

25. J. A. Eddy in Krupp, op. cit., pp. 133-163.

26. A. Ortiz, private communication.

27. J. H. Evans and H. Hillman, paper presented at the conference on Archaeoastronomy in the Americas, Santa Fe, N.M., June 1979.

28. We are indebted to Karl Kernberger for his excellent and invaluable photographic recording under difficult conditions. In many ways he and his assistants made possible the success of this project. We thank R. Blair, C. Hunt, A. Ortiz, and R. G. Vivian for their assistance on a number of important points. Conversations with W. Benson, M. Cohalan, L. E. Doggett, F. H. Ellis, G. Hawkins, P. Schaafsma, B. Wachter, and R. Williamson have been particularly helpful. We are grateful to W. Herriman and his staff at Chaco Canyon National Monument for their cooperation and permission to carry out the field work. Finally, we would never have completed this work, which has consumed our spare time for many months, without the patience and understanding of our families and friends.

2

Lunar Markings on Fajada Butte
Chaco Canyon, New Mexico

"Lunar Markings on Fajada Butte" appeared first in Archaeo-astronomy in the New World, A. F. Aveni, editor, published by Cambridge University Press in 1982. The late L. E. Doggett was an astronomer with the U.S. Naval Observatory, Washington, D.C.

F AJADA BUTTE IS KNOWN TO CONTAIN a solar marking site, probably constructed by ancient Pueblo Indians, that records the equinoxes and solstices (Sofaer *et al.* 1979a). Evidence is now presented that the site was also used to record the 18.6-year cycle of the lunar standstills.

Fajada Butte (Figure 1) rises to a height of 135 meters in Chaco Canyon, an arid valley of 13 kilometers in northwest New Mexico, that was the center of a complex society of pre-Columbian culture. Near the top of the southern exposure of the butte, three stone slabs, each 2 to 3 meters in height and about 1,000 kg in weight, lean against a cliff (Figures 2, 3).

Anna Sofaer
R. M. Sinclair
and
L. E. Doggett

FIGURE 1. *Fajada Butte from the north. The solar/lunar marking site is on the southeast summit.*

Behind the slabs two spiral petroglyphs are carved on the vertical cliff face. One spiral of 9½ turns is elliptical in shape, measuring 34 by 41 cm (Figure 4). To the upper left of that spiral is a smaller spiral of 2½ turns, measuring 9 by 13 cm (Figure 4).

SOLAR MARKINGS

Throughout the year near midday the two openings between the slabs form vertical shafts of sunlight on the cliff face. The daily paths of these dagger-shaped patterns on the cliff change with the sun's declination. As the patterns intersect the spirals, the equinoxes and solstices are uniquely marked.

At summer solstice one dagger of light descends through the center of the large spiral (Figure 4). On succeeding days the dagger descends discernibly to the right of center. As summer progresses and the sun's declination decreases, the position of the dagger shifts progressively rightward across the large spiral and a second dagger of light appears to the left.

At the autumnal equinox this second dagger bisects the smaller spiral (Figure 4). From fall towards winter the daggers continue their rightward

FIGURE 2. *The slabs from the south.*

FIGURE 3. *Eastern slab casting a sunrise shadow on the spiral. The inset shows the shadow of sunrise at declination +18.4° (i.e., northern minor standstill).*

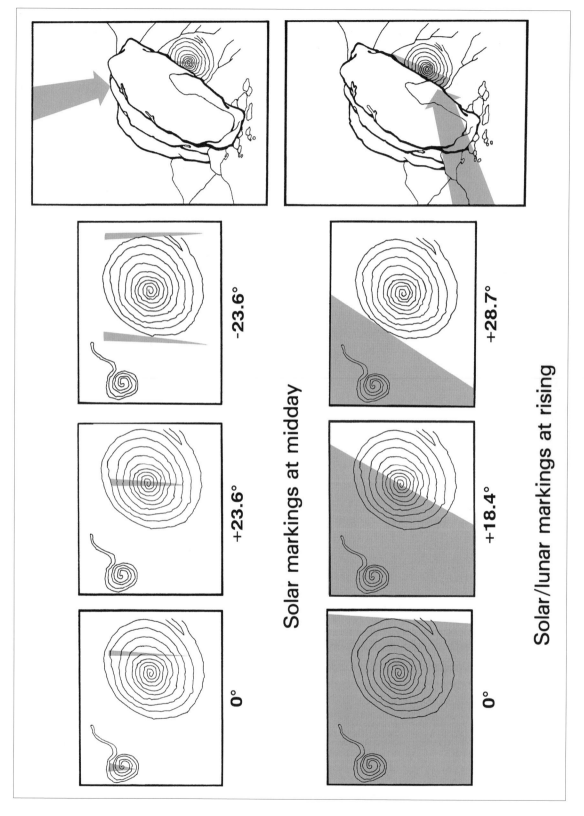

FIGURE 4. *Solar markings at midday and solar/lunar markings at rising at the indicated declinations.*

movement, until at winter solstice the two daggers bracket the large spiral, holding it empty of light (Figure 4). Following the winter solstice the cycle reverses itself until the next summer solstice.

In addition to the midday solar markings, we previously noted (Sofaer *et al.* 1979 b) that at sunrise the inner edge of the eastern slab casts a shadow on the larger spiral (Figure 3). The edge of this shadow crosses the spiral when the sun's declination is positive, and its position shifts leftward an average of 2.5 cm per week between the equinox and solstice (Figure 5). At the equinoxes the shadow edge falls in the far right groove of the spiral. Noting this second possible marking of equinox, we assumed that the sunrise shadows might form a second set of intentional solar markings. However, the locations of the shadows of the summer and winter solstices are of no particular note: at summer solstice the shadow edge falls between the center and the left edge of the spiral (Figure 5); at winter solstice the edge is not on the spiral at all. We noted the location of the shadow edge on the spiral as a sensitive indicator of changing solar declination for half the year, but with no certainty of its use or purpose.

Further observations through the year showed that when the sun is at declination +18.4° (mid-May and late July), the shadow at sunrise bisects the large spiral, putting the left half in shadow and the right half in light (Figures 4–6). At these times the sun is approximately at the declination of the moon at northern minor standstill. The edge of this shadow is aligned with a pecked groove (Sofaer *et al.* 1979a, Figure 7) that runs from the spiral's center to the lower left edge, emphasizing this particular occurrence. A further stimulus to search for lunar significance within this assembly was provided by Alfonzo Ortiz (1979), who suggested that because of the dual roles of sun and moon in Pueblo culture, a site in which the sun was so clearly marked would also include the moon.

LUNAR MARKINGS

Just as the rising sun casts a shadow, so too does the rising moon, provided the moon is in the proper portion of its monthly cycle of phases. Thus under the correct conditions (see Construction of the Site) the rising moon at minor standstill casts a shadow bisecting the large spiral.

While the sun provided a convenient simulation of moonrises up to declination +23.4°, we had to use artificial light sources to simulate the moon at higher declinations. (The next major standstill will not occur until 1987.) We used a laser for accurate alignment and near-parallel light from a floodlight to form the shadows. A series of simulations was calibrated against observations of various sunrises, with corrections made for the obliquity of the ecliptic of A.D. 1000, the estimated era of construction of the site. Also taken into account were the effects of lunar parallax and atmospheric refraction (Thom 1971) and the height of the eastern and northeastern horizon.

The simulation of the northern major standstill at declination +28.7° (epoch A.D. 1000) showed a shadow falling tangential to the far left edge of the spiral (Figures 4, 5).

We conclude that the Pueblo Indians recorded the extreme northern rising positions of the moon at major and minor standstills. In addition, as we speculated earlier (Sofaer *et al.* 1979a), the number of grooves in the spiral (counting horizontally from the left edge to the right edge) may record the length of the cycle. This appears in two possible ways: (1) as the cycle moves from minor to major standstill over 9 to 10 years, the extreme position of the lunar shadow shifts over the 10 grooves on the left side of the spiral; (2) the length of the full cycle (18.6 years) may be recorded by the count of 19 grooves across the full spiral. The number of grooves may also record a knowledge of the 19-year Metonic cycle. In addition the passage of the shadow edge through the far right groove of the spiral may record the midpoint of the declination cycles of the sun and moon.

The following factors can affect the position of the edge of the shadow for a given sunrise or moonrise (Thom 1971): variations in atmospheric refraction (±0.4 cm), the lunar wobble of 9' amplitude and 173-day period (±0.3 cm), and variations in lunar parallax (±0.1 cm). None of

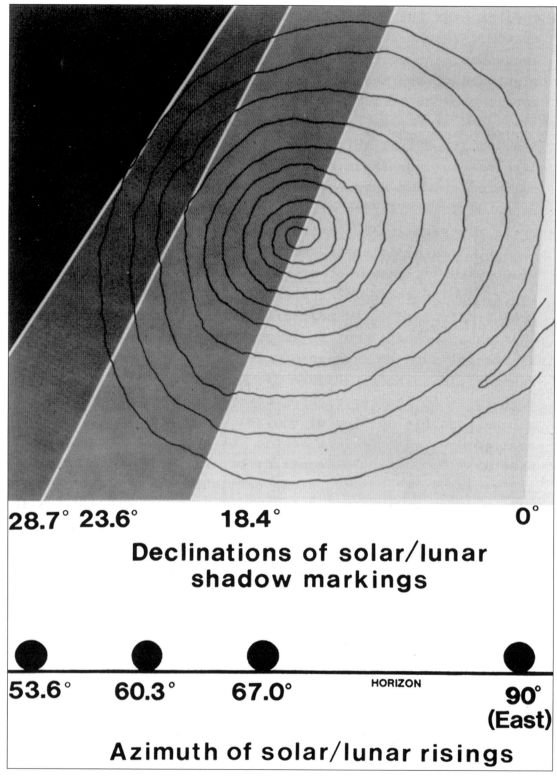

FIGURE 5. *Solar/lunar markings and solar/lunar risings on the northeast and eastern horizon.*

these introduces an appreciable uncertainty in the display.

In addition it is not clear what position of the rising moon would have been used by the ancient Pueblos to observe shadows. The difference in the shadow position on the spiral of two reasonable possibilities—the lower limb just tangent to the horizon and the lower limb positioned one lunar diameter above the horizon—is only 0.6 cm. The phase of the moon does not affect this, as long as a consistent definition of moonrise is used.

CONSTRUCTION OF THE SITE

We previously presented evidence (Sofaer *et al.* 1979a) that the slabs were deliberately placed and were possibly shaped on critical edges to form the midday patterns on the spiral. In doing so the builders had control over the placement of the edge that casts shadows at sunrise and moonrise. Calibration of this edge could have required shaping of the inner surface of the eastern slab (Figure 3). Further examination of this edge is required to determine whether its shape is natural or artificial.

Prior knowledge of the standstill cycle would most likely be necessary to relate the shadow casting edge in proper orientation to the spiral. This knowledge could most easily have been gained by accurate horizon watching, a practice reported by McCluskey elsewhere in these Proceedings to have been extensive among an historic Pueblo group, the Hopi.

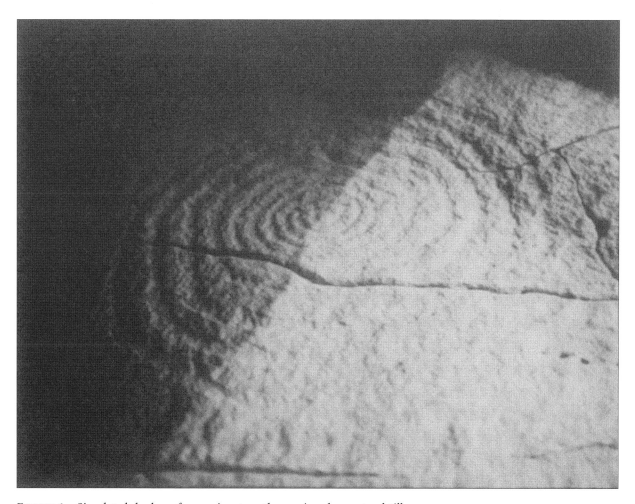

FIGURE 6. *Simulated shadow of moonrise at northern minor lunar standstill.*

Under clear weather conditions a majority of moonrises and moonsets are visible each month, with the shifting azimuthal position revealing first the moon's monthly cycle in declination and then, over the years, the standstill cycle. At Chaco Canyon the monthly declination cycle causes the azimuth of moonrise to vary from 67° to 113° near a time of minor standstill. These limits gradually increase over 9.3 years until at major standstill the azimuth varies from 54° to 126°. On the western horizon the setting cycles mirror the rising cycles on the eastern horizon.

Near a standstill the azimuthal limits change very slowly, making it difficult to pinpoint the exact year in which a standstill occurs. However, this makes possible a reasonably accurate determination of the amplitude of the standstill cycle, even if some potential observations are lost because of bad weather or the moon being at the wrong phase. With a desert environment of clear skies and an unpolluted atmosphere, the Pueblos had the advantage of optimal observing conditions.

Once the standstill cycle was known and the device constructed, the northern limits would be indicated on the spiral by appropriate moonrises that are bright enough to cast shadows. In New Mexico we have observed lunar shadows at moonrise, within a few minutes after the lower limb is tangent to the horizon, from the night after full moon to a few nights after third quarter. Since the moon's phase cycle (29.5 days) is longer than its monthly cycle in declination (27.3 days), the phase at successive declination maxima slowly changes. Thus in any give year there are approximately four

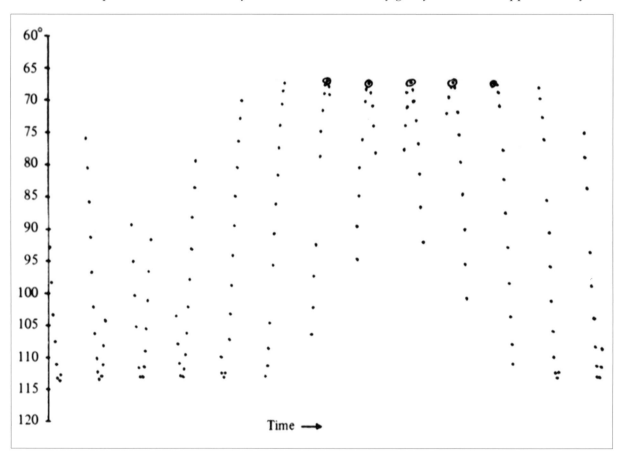

FIGURE 7. *Azimuths of moonrises occurring after the end of evening civil twilight and before the beginning of morning civil twilight during the year of a minor standstill. Moonrises with declination within 1° of the standstill limit and with sufficient brightness to cast shadows are circled.*

shadow casting moonrises occurring near the northern limit for that year (Figure 7).

CULTURAL BACKGROUND

The cultural and technological sophistication of the ancient Pueblo Indians of Chaco is evident in their development of an extensive trade and road network and in their planning and building of elaborate multi-story pueblos (Hayes *et al.* 1981). Interest in astronomical orientation is found in the reported (Williamson *et al.* 1975) solar and cardinal alignment of several pueblos and kivas (the Pueblo ceremonial structures in Chaco Canyon). It is also interesting to note the possibility of cultural contact with the Mesoamerican societies that had studied eclipse cycles (Lounsbury 1978) and developed complex calendric systems.

In the absence of direct knowledge of the customs of the prehistoric Pueblos, we turn to the historic Pueblos for insights into the ceremonial importance given to bringing together the cycles of the sun and moon. Many ethnographic reports of the scheduling of the winter solstice ceremony indicate strong desire to have the date coincide with the full moon (Stevenson 1904; Bunzel 1932; Ellis 1975). McCluskey (1977) reported that the Hopi synchronized the lunar and solar cycles over two to three years in setting their ceremonial calendar. More recently McCluskey (1981) has suggested that the Hopis' attention to the moon must have brought them close to observing the standstill cycle: "It would have been a short step for them to look for the moon's house, the theoretical lunistice which the moon reaches every 18.6 years..."

Spier (1955) reports that common to most of the historic Pueblos is the starting of the new year with the new lunation closest to winter solstice. Frequent planting of prayer flags at full moon, especially at winter solstice, also indicates the moon's significance in the Pueblos' ritual life (Bunzel 1932). The duality theme in Pueblo cosmology links sun and moon as male/female: sun-father and moon-consort or sister (Stevenson 1904). Ortiz (1981) reports the Tewa Pueblo group as seeing the moon as the mask of the sun.

There was thus a consistent effort to seek the synchronization of lunar and solar cycles. We speculate that there was success in this quest on Fajada Butte in bringing together at the spiral's center and outer boundary the highest and lowest positions of sun and moon (Figure 4).

With the possible exception of Casa Grande (Evans & Hillman 1981), we know of no other evidence of markings of the lunar standstill cycle in the Americas. We have searched for other possible explanations for the timing of the shadow phenomena in the culture and weather patterns in Chaco. A scholar of Chaco's agrarian prehistory (Truell 1981) found nothing significant in the dates when the sun reaches declination +18.4°, and weather patterns of Chaco do not indicate these dates as consistent times of rain or other climatic events. And of course the marking of declination +28.7° is not relevant to the annual solar calendar. The evidence does point to this site as a place where ancient Pueblos integrated on one set of spirals with one set of slabs the cycles of the sun and moon.

ACKNOWLEDGEMENTS

We are indebted to A. Schawlow and T. Hansch for supplying the laser used in the simulations; to A. Aveni, J. Carlson, M. Cohalan, J. Eddy, F. Eggan, G. Hawkins and J. Young for many fruitful discussions; to K. Kernberger for photography; to J. McGrath and P. Wier for field work; to B. Jones, R. Nugent, T. Plimpton and R. White for moonrise shadow observations. Finally we wish to express our appreciation to W. Herriman, Superintendent, Chaco Canyon, for his essential help to this research.

47

REFERENCES

Bunzel, R. L.
1932. "Introduction to Zuñi Ceremonialism," in *47th Annual Report. Bureau of American Ethnology,* pp. 467–1086. Smithsonian Institution, Washington.

Ellis, F. H.
1975. "A Thousand Years of the Pueblo Sun-Moon-Star Calendar," in *Archaeoastronomy in Pre-Columbian America,* ed. A. F. Aveni, pp. 59–88. University of Texas Press, Austin.

Evans, J. H. and H. Hillman
1981. "Documentation of Some Lunar and Solar Events at Casa Grande," in *Archaeoastronomy in the Americas,* ed. R. A. Williamson, Ballena Press, Los Altos, Calif.

Hayes, A. C., D. M. Brugge, and W. J. Judge
1981. *Archaeological Surveys of Chaco Canyon.* National Park Service, Washington.

Lounsbury, F. G.
1978. "Maya Numeration, Computation, and Calendric Astronomy," in *Dictionary of Scientific Biography,* ed. C. C. Gillispie, Vol. XV, pp. 759–818. C. Scribner's Sons, New York.

McCluskey, S. C.
1977. "The Astronomy of the Hopi Indians," *Journal for the History of Astronomy,* 8, 174–195.

McCluskey, S. C.
1981. Comment on the paper "Stone Age Science in Britain," by A. Ellegard, *Current Anthropology,* 22, no. 2: 119.

Ortiz, A.
1979. Private communication.

Ortiz, A.
1981. Private communication.

Sofaer, A., V. Zinser and R.M. Sinclair
1979a. "A Unique Solar Marking Construct," *Science,* 206: 283–291.

Sofaer, A., V. Zinser and R.M. Sinclair
1979b. "A Unique Solar Marking Construct of the Ancient Pueblo Indians," *American Indian Rock Art,* 5: 115–125.

Spier, L.
1955. *Mohave Culture Items,* pp. 16–33. Northern Arizona Society of Science and Art, Flagstaff.

Stevenson, M. C.
1904. *The Zuñi Indians.* 23rd Annual Report. Bureau of American Ethnology, pp. 9–157. Smithsonian Institution, Washington.

Thom, A.
1971. *Megalithic Lunar Observatories.* Clarendon Press, Oxford.

Truell, M.
1981. Private communication.

Williamson, R. A., H. J. Fisher and D. O'Flynn
1975. "The Astronomical Record in Chaco Canyon, New Mexico," in *Archaeoastronomy in Pre-Columbian America,* ed. A. F. Aveni, pp. 33–43. University of Texas Press, Austin.

3

Astronomical Markings at Three Sites on Fajada Butte

This study was published in Astronomy and Ceremony in the Pre-historic Southwest, *John B. Carlson and W. James Judge, editors (Maxwell Museum of Anthropology, Anthropological Papers No. 2, University of New Mexico, Albuquerque, 1983).*

Anna P. Sofaer
and
Rolf M. Sinclair

A SEVEN-YEAR STUDY OF PREHISTORIC PUEBLO SITES of the Chaco culture has revealed that these people possessed a sophisticated astronomy. Evidence of this prehistoric astronomy consists of multiple light markings on petroglyphs and of several alignments of major structures. This paper presents the results of a recent study of 13 markings at three sites on Fajada Butte in Chaco Canyon, New Mexico; these results are presented in the context of earlier research in this area. Each marking is a distinctive pattern of shadow and light that appears on a petroglyph at a key point in the solar or lunar cycle, i.e., at an extreme or midposition of these cycles, including the meridian passage of the sun at solar noon. [Note: times quoted throughout this chapter are in apparent solar time, in which noon occurs each day when the sun is due south on the meridian. The basic astronomical concepts used here are explained in Aveni (1980) and Krupp (1977).] Many of the markings simultaneously record two key points in different cycles, such as noon and solstice or noon and equinox. A minimum of 17 key points are indicated by the markings at these three sites.

The markings define the outer boundaries and midpoints of the recurring cycles of the two most basic celestial bodies: the sun and the moon. The accuracy and redundancy of the markings of these symmetric points, as well as the strong visual effect of the markings themselves, indicate their intentional quality. Several of the markings record meridian passage of the sun within a few minutes, and one marks equinox to within a day. An accuracy equivalent to that of these markings is found in the cardinal alignments of major constructions in the area.

Fajada Butte stands 135 meters high at the south entrance of Chaco Canyon. The remains of various structures, including a small kiva, and the presence of potsherds and rock art on and near the butte summit show that it was frequented by prehistoric Pueblo people and by Navajo people. The content and style of the glyphs at the three sites reported here indicate that these markings are of prehistoric Pueblo origin and were probably made between A.D. 900 and 1300.

From about A.D. 900 to 1150 Chaco Canyon was the center of a complex prehistoric Pueblo society that thrived in the arid environment of northwestern New Mexico. The Chacoan people planned and constructed large multistory structures, ceremonial centers, and extensive roads throughout the 70,000 square kilometer San Juan Basin. These achievements indicate sophisticated engineering and surveying skills. Ethnographic reports concerning the historical Pueblos (some of whom are descendents of the Chacoan people) reveal their keen interest and skills in observing the sun and moon for ritual and agrarian purposes (see references cited in Sofaer *et al.* 1979).

The Fajada markings, although similar to others recently reported in the Southwest, are unique among known archaeoastronomical sites in several respects: (a) they use the changing altitude of late morning and midday sun to indicate the solstices and equinoxes, (b) they record solar noon on rock art, and (c) most especially, they combine recording of both the season and noon in the same markings. The markings at one site are the only known ones in the New World that combine recordings of the extremes and midpositions of the moon and the sun. The use of the features and topography of a prominent land mass to create these complex and varied markings on carved petroglyphs is also unique in our current knowledge of prehistoric astronomy.

Following a brief summary of earlier findings at one of the three sites near the top of Fajada Butte (see also Sofaer *et al.* 1979, 1982a), this chapter describes in detail the recently discovered markings that combine noon and seasonal recordings at two additional sites on the east and west sides of the butte (Sofaer *et al.* 1982b). New observations and analyses regarding the earlier findings are then presented. The sites are referred to in this paper as the three-slab site, the east site, and the west site. The markings of these sites were first observed and studied by Sofaer between 1978 and 1983. The markings are summarized in Table 4.1. Ethnographic and archaeological correspondences with the markings are also discussed.

SUMMARY OF EARLIER FINDINGS

At the three-slab site, six markings record solar and lunar positions (Figures 4.1, 4.2, 4.3). An unusual configuration of three large stone slabs, each about 2 meters high, collimates sunlight each day in the late morning and near midday onto two spiral petroglyphs pecked on a cliff face. The streaks of light so formed change noticeably with small changes in the sun's declination. The vertical light/shadow patterns on the petroglyphs thus go through an annual cycle in which the solstices and equinoxes are marked by the intersection of these patterns with the primary features of the spiral forms (Sofaer *et al.* 1979). This site also marks the northern minor and major extremes of the 18.6-year lunar standstill cycle by a separate pattern of light and shadow at moonrise (Sofaer *et al.* 1982a). This light and shadow pattern marks events that are unrelated to those of the solar cycle, yet it is also formed on the primary features of the larger spiral by another edge of one of the three slabs that form the midday patterns.

Since the moon can appear throughout the declination range of the sun, all the solar markings listed in Table 4.1 can also be formed at certain times by the moon. We do not consider most of these to be possible lunar markings owing to a lack of further evidence. The northern minor standstill marking can be formed at certain sunrises away from the solstices and equinoxes; we term it lunar because it occurs as a pair with the northern major standstill marking (which cannot be solar since the moon then exceeds the maximum solar declination). The equinox sunrise marking (Figure 4.1) also indicates the middle of the moon's monthly or 18.6-year declination cycles, in a manner analogous to the markings of the standstill extremes, and is thus possibly also a lunar marking.

TABLE 4.1. *Markings of astronomical cycles on Fajada Butte*

Petroglyph	Solar					Lunar	
	Midday				Sunrise	Moonrise	
	Noon	Summer Solstice	Equinox	Winter Solstice	Equinox	Northern Major Standstill	Northern Minor Standstill
Large spiral (three-slab site)	–	X	–	X	X	X	X
Small spiral (three-slab site)	–	#	X	–	–	–	–
East spiral	X	X	X	X	–	–	–
East snake	X	–	X	–	–	–	–
East rectangle	X	X	X	–	–	–	–
West double spiral	X	–	X	–	–	–	–
West rectangle	X	–	–	–	–	–	–

The <u>absence</u> of light at this location can be taken as a marking (see text)

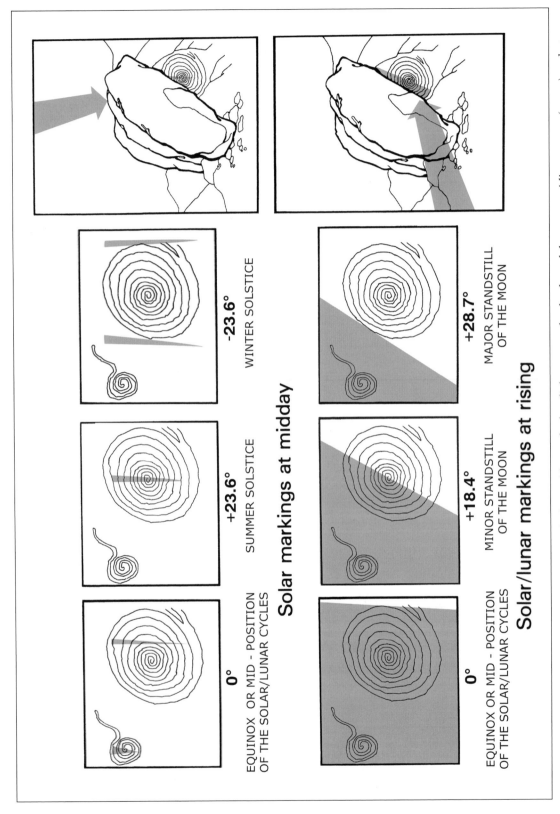

Solar markings at midday

Solar/lunar markings at rising

FIGURE 4.1. *The three-slab site. Right: Formation of solar and lunar shadow/light patterns by the three slabs near meridian passage (upper) and at rising (lower). Left: Schematic of the resulting patterns on the spirals at the indicated declinations and seasons (see Sofaer et al. 1979, 1982a for detailed photographs). (Note: an error in an earlier publication [Sofaer et al. 1982a: Figure 4] of "rising 0°" is corrected here.) (Illustration by Pat Kenny; Copyright © Solstice Project.)*

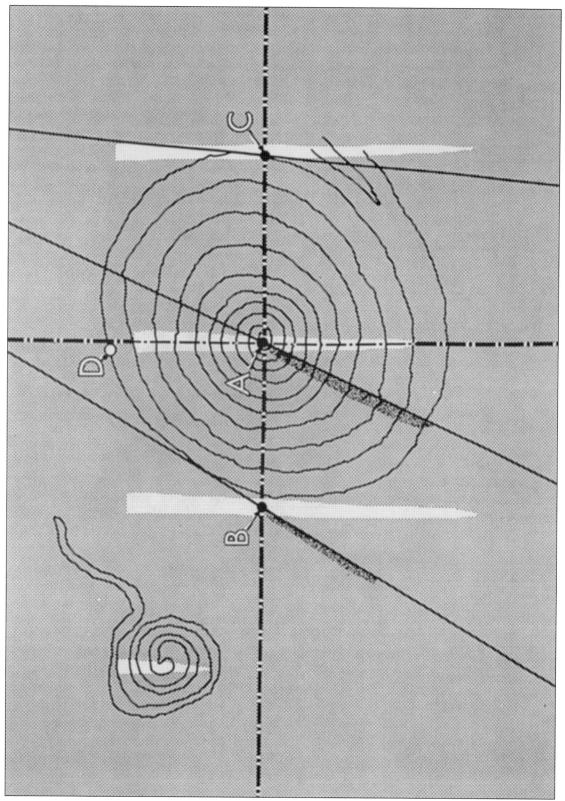

FIGURE 4.2. *The three-slab site. Superposition of the main elements of the six markings (see* FIGURE 4.1*) showing the multiple uses of certain key features of the large spiral (A), points of tangency at left and right (B and C), the top (D), and the horizontal and vertical axes. Also shown are the two pecked grooves (see text). (Illustration by Pat Kenny, Copyright © Solstice Project.)*

While the solar markings of the three-slab site use the seasonal altitude changes of the late-morning and midday sun to indicate the specific time of the solstices and equinoxes, they do not accurately mark noon. New evidence reveals that two other sites on the butte mark the specific time of noon at the seasonal points. Several features of the three-slab site are shared by the two other sites: spirals, dagger-shaped light patterns, and the repeated use of the same glyph or pair of glyphs at quarter points of the seasons.

RECENT FINDINGS
OF NOON/SEASONAL MARKINGS

At two sites located a short distance below the three-slab site (Figure 4.4), five petroglyphs are crossed by visually compelling patterns of shadow and light at a time close to solar noon. The imagery of these patterns distinguishes the solstices and equinoxes in most instances. These shadow patterns form seven markings indicating 11 key points in the daily and seasonal cycles of the sun — i.e., midpositions and extremes (Table 4.1). The markings occur in combinations, most of them in striking conjunctions: pairs that are visible at the same site and that vary with the seasons.

The shadows that form the noon markings are cast by rock edges of the butte, which is locally very irregular. Since the sun's elevation at meridian passage changes annually from 31 degrees at winter solstice to 78 degrees at summer solstice, the particular edges that cast shadows on the glyphs through the year differ greatly in distance and bearing from them. Some edges are less than a meter away, while others are up to 30 meters distant. During the sun's daily meridian passage, the glyphs at the east site change from fully lit in morning sun to fully shadowed in the afternoon, and those at the west site go through the reverse transition of shadow to light.

At the east site (Figures 4.5 and 4.6), which is located about 25 meters below the butte summit, three adjacent glyphs—a nearly vertical rattlesnake (22 cm long), a rectangular figure (14 cm wide), and a spiral (15 cm wide) occur 2.0 to 2.5 m above the cliff base —the rattlesnake and the rectangular figure are particularly deeply incised. The spiral is pecked within a rectangular area that appears to have been worked. A shadow edge crosses the spiral glyph within ten minutes of noon throughout the year, forming a seasonally changing pattern. This pattern is momentarily symmetric about the center of the spiral within a few minutes of noon, forming a wedge at summer solstice, a quartering at equinox, and a bisecting at winter solstice.

The seasonal variation is reinforced by simultaneous markings that occur on the two nearby glyphs only within a few weeks of summer solstice and equinox, close to noon. At equinox, at the same time that the spiral is quartered, an unrelated shadow edge crosses the snake, touching all parts at once: the head, the body, and the rattles. This shadow pattern deviates noticeably from alignment with the full body of the snake two weeks earlier or later (Figure 4.6). At summer solstice, at the time that the wedge marks the center of the spiral, the shadow edge is aligned with the right edge of the rectangular figure (Figure 4.5). This combination is noticeably different two weeks before and after summer solstice. Thus, the seasonal points of the year are indicated both by the differing patterns on the individual glyphs and, at summer solstice and equinox, by the unique paired combinations of the markings.

At the west site (Figures 4.7-4.9), which is located about 30 m below the summit, a double spiral (55 cm across) and a rectangular figure (17 cm wide) are about 6 m and 4 m, respectively, above the current cliff base. On days near equinox, close to noon, a narrow dagger-like pattern of light moves in an upward diagonal course through the double spiral in about 22 minutes. An unrelated vertical shadow edge moves across the rectangle in nearly the same interval. At equinox the light pattern moves through the center of the right whorl of the double spiral, with the whorl and the rectangle being bisected simultaneously about nine minutes before noon. The moving light patterns reach the far right edges of both glyphs within a few minutes of noon.

FIGURE 4.3. *Summer solstice light marking on the larger spiral of the three-slab site. The four arrows indicate the two pecked grooves with which the lunar shadow patterns are aligned. (Photograph by David Brill, Copyright © 1982 The National Geographic Society.)*

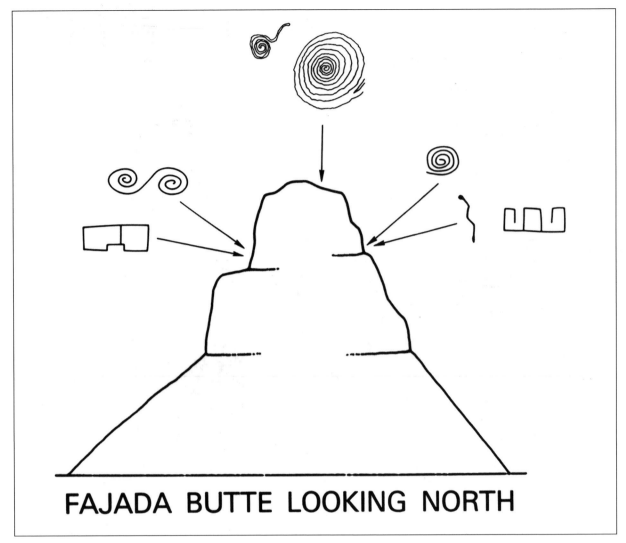

FIGURE 4.4. *Schematic of Fajada Butte (looking north) showing the locations of the petroglyphs at the east, west, and three-slab sites (which are not intervisible). (Illustration by Pat Kenny, Copyright © Solstice Project.)*

Movement of light across the double spiral near equinox is a sensitive indicator of the sun's declination (which is then changing most rapidly). The track of the pattern shifts 4.5 cm each day, left to right in the spring and right to left in the autumn, so that the pattern of light movement for each day is quite distinct during this period. Figure 4.9 shows the light pattern on several dates near equinox. In the photographs of the double spiral near equinox, the light form is shown at the same height in its upward diagonal course for each day. The "event" of equinox can occur through the years at different times on the nominal "day" of equinox. Figures 4.7 and 4.8 show that a shift in successive years of only 12 hours in the time of equinox can be seen. By noting the exact position of the light pattern with respect to the center of the right whorl, one can pinpoint equinox to within at least a half day.

The shift of the track of this light pattern from left to right on the double spiral could be used to anticipate the approach of spring equinox [Chapter 3], since even to a casual observer a one- or two-day shift from equinox would be evident. While no

56

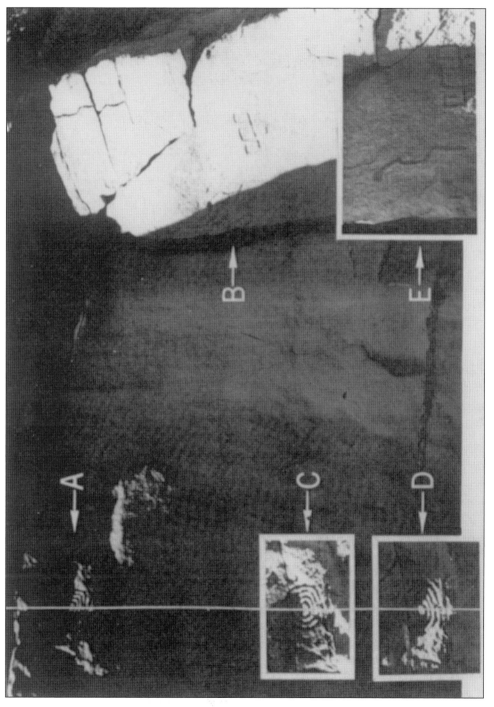

FIGURE 4.5. *East site. Simultaneous markings at equinox on (A) spiral and (B) snake (September 20, 1979, 12:04 pm). Insets: (C) winter solstice marking on spiral (December 20, 1980, 12:00:30 pm), (D and E) summer solstice simultaneous markings on spiral and rectangular figure (June 22, 1978, 12:03 pm). The line through ACD points out the change in the shape of the shadows that centrally mark the spirals at each season around the time of noon. (Photograph by A. Sofaer and K. Kernberger, Copyright © Solstice Project.)*

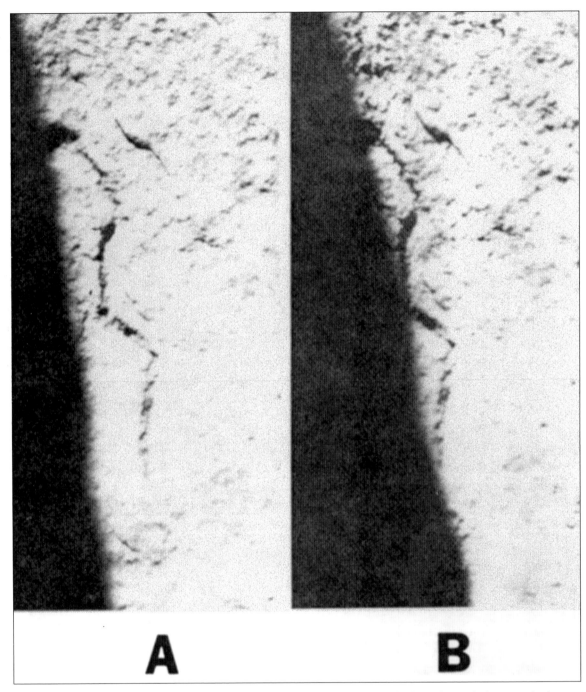

FIGURE 4.6. *East site. Markings on the snake showing the alignment of the shadow edge with the petroglyph at equinox. (A) 16 days after fall equinox (October 9, 1982, 12:00 noon), (B) 3 days before fall equinox (September 20, 1979, 2:03:45 pm). [Note: data taken a certain time after (or before) spring equinox are equivalent to data taken the same time before (or after) fall equinox.] (Photographs by A. Sofaer, Copyright © Solstice Project.)*

external evidence is available to determine whether this site was used to mark the time of equinox so accurately, it was clearly possible.

The markings on the rectangle are of special interest in that they indicate noon accurately throughout the year. Each day a vertical shadow edge crosses the rectangle within 5 to 20 minutes of noon and aligns momentarily with the right edge within three minutes of noon throughout the year.

Figure 4.10 illustrates the intervals around noon between first and last touchings by a shadow edge on each glyph at both the east and west sites. This figure also shows certain times within these intervals when several of the glyphs are marked by distinctive and centrally aligned patterns; these are illustrated in Figures 4.5 to 4.9 as well. Several of these patterns occur within a few minutes of noon.

Several factors underscore the intentional quality of the markings described here. First, in all instances the light patterns bisect or align with the edges of the glyphs at key points in the solar cycles. Second, six of the seven markings record two such events (noon and a solstice or equinox). Third, the visual effect is striking, most especially when two symmetric shadow alignments occur simultaneously on two nearby glyphs. Fourth, six of the seven markings occur in such simultaneous pairs and are formed by unrelated shadows cast by different parts of the butte. Fifth, three markings use the same glyph in different patterns at different seasons. And sixth, four of the markings bisect or involve the centers of spiral glyphs (in a manner similar to markings at the three-slab site).

The rock edges casting the pattern on the rectangular figure at the west site in winter are vertical and about three meters away; in summer they are horizontal and about 30 meters distant. It is remarkable that a point was found where the shadow patterns cast by such a wide range of the butte's irregular surfaces would remain vertical throughout the year and cross the glyph consistently at noon.

The glyphs are so placed that no other shadow crossings occur near noon except those that form the markings; in general the glyphs are either steadily illuminated or shadowed at other times. The markings occur at the universally recognized extremes or midpoints in the annual solar cycle, as do those at the three-slab site. There are no equivalent markings at these sites that occur at times other than the key points in the solar cycle.

The astronomical markings involve a significant fraction of the known rock art on Fajada Butte. There are abut 20 clearly formed glyphs at various locations on the butte, primarily scattered over its uppermost one-third (Sofaer and Crotty 1977), and some 20 other amorphous peckings and scratches and historical graffiti. When the five glyphs that are involved in the east and west site noon/seasonal markings are added to the two glyphs at the three-slab site, the marked glyphs constitute about one-third of the Fajada Butte rock art. (The other two-thirds appear not to be astronomically marked.) At each site the marked glyphs are dominant in clarity, size, and number over the few other nearby ones. Four of the seven spiral glyphs on the butte are used in astronomical markings, as are the only two rectangular figures.

Although the markings are referred to here as occurring near noon, it is likely that they were designed to commemorate both a spatial and a temporal halving of the day by recording when the sun is due south. (Note that it is more accurate to determine meridian passage or noon by observing the sun's changing azimuth around a north-south line than by monitoring changes in altitude.)

Central to these accurate noon/seasonal markings are several important astronomical concepts and their implementation. First, the creators of the markings singled out for emphasis midday and the year's quarter points. Then they had to establish the north-south meridian to the necessary accuracy, note the sun's transit, and identify the specific times of the equinoxes and the solstices. With this information they examined the shadow transitions on the Fajada cliffs throughout the year for possible sites. The weathering of the sandstone edges that form the markings has made it difficult to determine whether the developers of the markings worked these edges or used them in their natural state.

FIGURE 4.7. *West site, 1.5 hours before fall equinox (September 22, 1984, 11:50:30 am). The shadow edge moves across the rectangular glyph (lower right) from left to right, crossing the right edge within three minutes of noon throughout the year. (The outline of this glyph is artificially emphasized in this illustration.) (Photographs by Colin Franklin, Copyright © Solstice Project.)*

We conclude that the three Fajada sites were made to exhibit the extremes and midpoints of basic astronomical cycles considered significant by the Chacoan people. The markings occur only in a narrow band of times of day and year that is centered on astronomically recognizable events. An argument could be made that the petroglyphs were originally placed with no regard for their illumination and that shadows could randomly cross some of them at these times (particularly when the sun and shadows are moving fastest). This argument, however, cannot easily explain the pairings of markings, the multiple use of certain carvings, the repetitions of the alignment of two simple shadow patterns with the basic geometry of the petroglyphs at just the times of astronomical significance, or the consistent use of the spiral motif.

FURTHER OBSERVATIONS
AND ANALYSIS OF THE THREE-SLAB SITE

Lunar Markings. New evidence and further analyses of this site support earlier findings that the northern minor and major lunar standstills are marked at moon rising and refute a speculation that the moon's meridian passage is marked.

The large spiral glyph at the three-slab site contains a total of nine and one-half turns, making it unique among the numerous recorded rock art figures in Chaco Canyon. No other spirals recorded in the canyon have so many turns; most have no more than half as many. This fact strengthens the suggestion made earlier (Sofaer *et al.* 1982a: 176) that this number was chosen for the spiral to represent the number of years in the lunar standstill cycle. During the period of 9–10 years between the minor and major standstills, the position of the shadow cast by the most northerly rising moon each year shifts gradually across the nine and one-half turns of the spiral.

Further examination of the site has disclosed a pecked groove tangential to the large spiral and aligned with the major standstill shadow marking. This groove is similar to the pecked groove aligned with the minor standstill shadow marking (Figure

4.3; Sofaer *et al.* 1982a:173); these alignments draw attention to the shadow patterns as markings (Figures 4.1, 4.2).

It has been reported (Sofaer *et al.* 1982a: 178) that a standstill could potentially be marked at four moonrises in the years of the extremes. Further analysis shows that the criteria can be met by as many as 13 risings in some of these years (Michael Zeilik, personal communication 1981). This increased frequency would make the marking of the standstills that much easier to conceive of and achieve.

It was speculated earlier that a possible lunar marking could occur at the moon's meridian passage at the declination of the major standstill, when the moon achieves an altitude of about 82 degrees (Sofaer *et al.* 1979: 290). Recent measurements at the site show that an overhanging cliff edge about ten meters above the slabs would block the rays of the moon at this altitude so that no such marking could occur.

It is likely that the ancient Chacoans noticed the 18.6-year lunar cycle in the course of making horizon observations. During the 9–10 years between the standstills there is at Chaco a 13-degree shift in the extreme northerly and southerly positions of moon risings and settings; this shift is centered around the winter and summer solstice positions. The lunar extreme positions remain close to each standstill limit for a year or more. (The lunar standstill cycle would also be evident as the moon stood higher or lower in the sky in successive years and in the correspondingly changing nocturnal shadow and light patterns in the canyon and around and in the buildings.) In Chaco, the open, flat horizons and the clarity of the desert air of a thousand years ago would have made such observations far from difficult, especially for a people skilled in surveying. These rhythms of lunar cycles would become quite evident after some years of observation. The creators of the Fajada lunar markings did not necessarily know the standstill cycle to any better accuracy than approximately a year, which is what can be perceived in the movements of the shadows on the large spiral of the three-slab site.

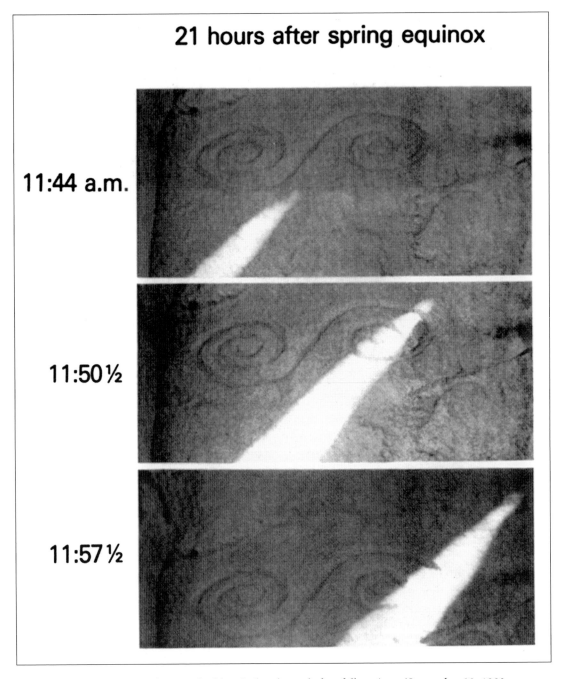

FIGURE 4.8. *West site. Marking on double spiral 21 hours before fall equinox (September 22, 1983, top to bottom: 11:44 am, 11:50:30 am, 11:57:30 am). (Photographs by Peggy Wier, Copyright © Solstice Project.)*

FIGURE 4.9. *West site. Marking on double spiral near equinox. Top: September 30, 1984, 11:42:30 am (8 days after fall equinox). Middle: March 20, 1984, 11:48:30 am (9 hours after spring equinox). Bottom: March 23, 1984, 11:53:30 am (3 days after spring equinox). (Photographs by Rolf Sinclair and Michael Marshall, Copyright © Solstice Project.)*

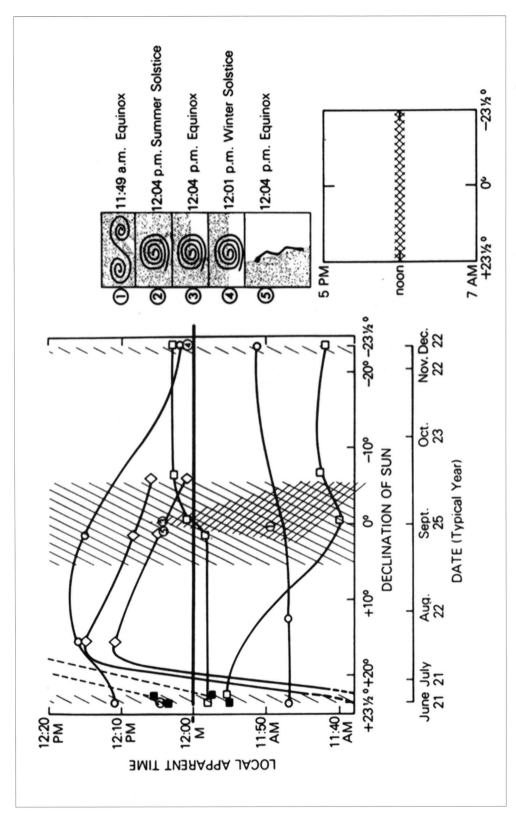

FIGURE 4.10. *Left: Time that shadow first touches a glyph and then just covers it (bottom and top curves of each pair). East site:* ■ *square,* ◇ *snake,* ○ *spiral. West site:* □ *square,* \\\\ *range of marking on double spiral.* //// *are the intervals of one month centered on solstices and equinox.*
Right: Moments near noon when shadow edges are centrally aligned on certain glyphs. These cases are located by numbers on the graph (left). (1) West site double spiral, March 20, 1984. East site spiral: (2) June 22, 1978; (3) September 20, 1979; (4) December 20, 1979. (5) Snake, September 20, 1979. Inset: Replotting of graph on left to show the fraction of daylight hours involved in the markings. (Illustration by Pat Kenny; Copyright © Solstice Project.)

These lunar markings do not necessarily signify that the Chaco culture was predicting lunar eclipses. The 18.6-year cycle is of limited use for this purpose (see Chapter 2 in *Astronomy and Ceremony in the Prehistoric Southwest*, J. B. Carlson and W. J. Judge, editors). It is worth noting, in this regard, that the modern Pueblo cultures have shown far more interest in cycles than in unusual events (Ellis 1975: 60–63). The extremes and midpositions of recurring cycles are themselves significant because they define the limits and order of the cosmos. Recognition of these cycles provides reason enough to commemorate them.

INTERRELATIONSHIPS OF THE MARKINGS AND THE ASSEMBLY OF THE THREE-SLAB SITE

The three-slab site combines a number of markings in a unique manner. New analyses presented here provide insights into its design and development.

Because of the extensive seasonal altitude changes of the midday and morning sun and the angles of the slabs relative to the cliff face, many different portions of the slabs create different parts of the complex set of markings. Nine separate surfaces are involved in casting the shadows that form six markings of five different declinations. (Note a possible seventh marking: Light is almost entirely blocked by the left and middle slabs at summer solstice [Sofaer *et al.* 1979: 286]; only 10 days later a streak of light is much more evident and then grows through the months to become the bisecting equinox pattern. The effect at summer solstice may have been intended to focus attention solely on the bisecting pattern on the large spiral [Table 4.1].)

Each of the five solar and lunar markings on the large spiral interlocks with the positions of the other markings, in that each of these markings falls on the points and lines of symmetry that define the large spiral (Figure 4.2). There are five points that define the spiral: center, left edge, right edge, top, and bottom. The first four of these points are clearly part of the astronomical patterns. At the extremes and midpositions of the solar and lunar cycles a pattern of shadow and light strikes one and sometimes two of them. Conversely, most of these four points are involved in two markings. To a lesser extent the fifth point, the bottom edge of the spiral, is also defined by the markings.

At summer solstice the bisecting dagger-shaped pattern is centered simultaneously on both the vertical and horizontal axes of the spiral (Figure 4.1). The minor lunar standstill marking, which is formed by a different rock edge, also crosses the center of the large spiral.

Three shadow markings are tangent to the right and left edges of the large spiral. These are the winter solstice pattern, the lunar major standstill marking, and the lunar/solar 0-degree declination marking (Figure 4.1). These markings emphasize the right and left edges of the spiral.

The top edge of the large spiral is defined by another aspect of the markings. At summer solstice the dagger-shaped pattern starts as a dot of light on the top turn of the spiral. Only ten days earlier or later this effect is lost: the first light on the cliff face is 10 cm above the spiral. The bottom edge of the spiral is related to, and hence perhaps defined by, the momentary symmetrical positioning (mentioned above) of the summer solstice light dagger.

It is possible to arrive at some estimates of the degree of sensitivity in the relationships among the rock slabs and the markings. Because of the oblique angle between the slabs and the cliff face, moving the rock edge that casts the moonrise shadows 1 cm to the north or south would create twice this displacement in the positions of the lunar shadows on the spiral and thus discernibly shift the markings. A movement of either the eastern or middle slab by 2 to 3 cm to the east or west would displace the shadows that form the midday light patterns by that much. The summer solstice pattern is only 2 cm wide, so such a change could block the light entirely. Because each of the slabs forms a part of up to five markings, any such shift would change and probably destroy more than one marking.

A recent report (Newman *et al.* 1982) presents a scenario in which the rocks could have fallen naturally close to their present positions. The evidence presented in that report does not, however, exclude the possibility of later deliberate movement of the rocks to create the markings. Indications of possible shaping of the slabs and the cliff face where the spirals were pecked are discussed in an earlier publication (Sofaer *et al.* 1979: 289).

Both the interlocking nature of the markings and their sensitivity to the exact positions and shapes of the rock slabs indicate that moving and shaping of the slabs very likely took place. A few people could have moved and adjusted the slabs in small increments by simple techniques. Such manipulation would have made it easier to attain the interrelated markings. The complexity of the markings suggests that extensive planning, observation, and experimentation were required to achieve them.

The technology for shaping, moving, and using large slabs was well established in Chaco Canyon (Hewett 1936: 87–88; Mindeleff 1891: 148) and at other places in this cultural area, and many of these slabs are significantly larger than those at the three-slab site (Hayden 1878: 429; Mindeleff 1891: 58, 147–148; Newcomb 1966: 137). Some isolated and implanted slabs are also known to have been used in historical times for astronomical purposes (Mindeleff 1891: 86, 148).

Caution is appropriate in assessing the degree of artificiality in structures of the Pueblo culture. The natural appearance of many historical and prehistoric sites, especially shrines, conceals elements of construction and may be intended for the purposes of protective secrecy and integration with nature (Alfonso Ortiz, personal communication 1983). Stevenson's description of a shrine at Zia Pueblo illustrates the subtlety of construction at these sites:

> [A] stone slab rested so naturally on the hillside that it had every appearance of having been placed there by other than human agency. The removal of the slab exposed two vases side by side in a shallow cave [Stevenson 1894: 90].

PREHISTORIC PARALLELS TO THE FAJADA MARKINGS

Shadow-light formations mark critical times in the solar cycle at many prehistoric sites in the Chaco cultural region and elsewhere in the world. Many share the motifs of the Fajada markings: dagger-like shapes of light, spirals, snakes, and vertical shafts channeling light patterns at noon.

It is reported that winter solstice is marked when the rising sun's rays are collimated by a window in Pueblo Bonito, a major structure in Chaco Canyon (Reyman 1976). At Hovenweep, in southeastern Utah, an ancient Pueblo community lying just to the northwest of the Chaco region, several light markings are reported (Williamson 1981: 68–70). At the solstices and equinoxes light is collimated by portholes to fall in the corners and midpoints of tower structures. At a nearby rock art site two light daggers bisect two concentric circles and a spiral at summer solstice sunrise. Shadow/light formations mark these glyphs again at equinox sunrise, at which time they also mark a snakelike form. Recent reports of solstice and equinox markings at an eastern Arizona site near the Chaco region include a great number of bisecting dagger shapes of light on rock art, often on spirals (Carlson and Judge, *Astronomy and Ceremony in the Prehistoric Southwest,* chapter 11). Similar observations are reported in California (Krupp 1983: 129–137).

A number of the sites of monumental architecture in Mesoamerica have been noted for their symbolic display of shadow and light (Aveni 1980: 284–286). For example, a snake-shaped shadow is formed at equinox along the edge of a Mayan pyramid at Chichen Itza. While this marking is several hundredfold larger than the small snake marked by a shadow on Fajada's east site, the parallel is curious: both sites record equinox with shadow alignments on nearly vertical snakes. At two other sites in Mesoamerica the sun's rays at zenith passage are channeled through vertical shafts to fall as discs of light on the subterranean floors of the structures (Aveni 1980: 253–256).

Despite these parallels the Fajada markings remain distinctive in at least three respects: their combining of noon and season, their combining of sun and moon, and their use of rock art and rock slabs in numerous markings clustered on a single, prominent butte. Some of the particular characteristics of the Fajada markings can be understood in the context of historical Pueblo culture.

ETHNOGRAPHIC BACKGROUND

Ethnographic accounts of the historical Pueblo cultures describe concepts fundamental to the Pueblo people's world view. These concepts often have significant parallels among findings about prehistoric Pueblo communities.

Some discontinuity between prehistoric and historical Pueblo culture over the intervening 800 to 1000 years should be expected as a result of environmental and cultural change and migration, including the Spanish entrada in the 1500s (Berry 1982; Cordell 1984; Upham 1984). Certain gaps exist in the ethnographic record of astronomical practices because the Pueblo traditions of secrecy protect all ceremonial activities, including solar and lunar observations.

> Everyone who has worked among Pueblo Indians realizes only too well how averse they are to revealing the details of this manner of life. This attitude on the part of native informants makes it virtually impossible to secure a complete record of any Pueblo tribe [Titiev 1944: 4].

Fewer than 20 of the 80 or more historical pueblos extant when the Spaniards first came survived to be included in ethnographic studies. Thus, a great deal of information about Pueblo culture is lost or not available. The information that does exist can be used to provide general insights into the significance of prehistoric findings rather than to verify or refute interpretations of specific phenomena.

The historical Pueblos conceived of complementary roles of sun and moon (see references cited in Sofaer et al. 1982a) and used both of them intensively in timing of planting and ceremonial activity. These activities are consistent with the overlapping systems of marking of the solar and lunar cycles at the three-slab site. A recent study (Tedlock 1984: 6) describes clearly the complementary roles of sun and moon at Zuñi. In timing of the ceremonies for the solstices, the "weak light" of the winter solstice sun is matched with the "bright light" of the full moon, and the "bright light" of the summer solstice sun is matched with the "weak light" of the new moon. The location of the astronomical markings high on the seemingly remote and inaccessible Fajada Butte is in keeping with ethnographic reports of the historical cultures. Buttes, mesas, and mountains are often regarded as sacred places, and shrines are placed on their tops (Boas 1925-1928: 39–40; Dumarest 1919: 206–207; Ortiz 1969). Some shrines on high sites are used for solstice ceremonies (Kallestewa et al. 1984; Stevenson 1904: 109, 149).

Specific features of the complexly organized communities of the prehistoric Pueblos are not found or reported among the historical Pueblos. For example, the elaborately planned multistory architecture and engineered road system of the Chaco culture are not present among the historical Pueblos. Precise equinox markings and cardinal alignments are also not evident. The general period of the equinoxes is of major significance among the Tewa Pueblos, however (Ortiz 1969). Similarly, there are no reports in the ethnographic record of knowledge or markings of the 18.6-year lunar cycle among the historical Pueblo cultures, although there is a possible hint of previous knowledge of this cycle in the cosmogony of one pueblo (Stevenson 1894: 71).

There are numerous reports of Pueblo people observing shadow and light patterns cast by upright slabs or by windows and doorways on walls and floors of houses, kivas, and chief's houses to time ceremonial and agrarian activities (Cushing 1979: 117; Lange 1959: 56, 249; Mindeleff 1891: 86, 148). On ceremonial occasions light might fall on a quartz crystal, a bowl of water, a deer skin, or a kiva bench (Lowie n.d.; Parsons 1929: 176; Titiev 1944: 105).

Meridian passage is known to be significant in the cosmology of historical Pueblo people and is an integral part of many myths and ceremonies, often involving light markings. One such instance involves the winter solstice noon.

> In the roof of the ceremonial room there is a hole through which at noon the sun shines on a spot on the floor near where the chief stands....All sing the song of "pulling down the sun."...This is noon time when for a little while the Sun stands still [Parsons 1932: 292-293].

A similar ritual is practiced at summer solstice (Parsons 1932: 297). In keeping with the complementary roles of sun and moon, a report states, "The ritual of bringing down the moon seems to be much the same as that of bringing down the sun," and in this ritual the moon is said to stay "until noon" (Parsons 1932: 330). Another study reports that the moon's meridian passage is noted in the timing of Pueblo ceremony (Tedlock 1984: 94, 108).

Several other ritual practices and traditions of the Pueblo culture convey the significance of noon as the time when the sun stands still or rests for a short time in the middle of its course through the day (Boas 1925–1928: 284; Dumarest 1919: 222) and as the time when "he stops for dinner" (Curtis 1926: 104; Dumarest 1919: 222). The cacique conducts ceremonies in which the sun as a disc is moved across a screen in the arc of its day's course; he orders the sun at the middle of its course to stop for a short while at its noon position (Dumarest 1919: 198; Lange 1959: 267).

In one pan-Puebloan tradition, the Sun impregnates a virgin and she gives birth to the son(s) of the Sun (Parmentier 1979: 609). The impregnation and the birth often occur at noon and sometimes near summer solstice (Benedict 1931: 31, 1935: 46; Cushing 1931: 429, 431, 436; Dumarest 1919: 217; Gunn 1917: 129; Parsons 1932: 393, 1940: 55–56; Stephen 1929: 11–14). In several versions of this tradition the sun's rays penetrate a window or hatchway of a dwelling and fall on the lap of the maiden. In one version the virgin (yellow woman of the north) stops to rest on her journey to the center of the earth, and the Sun when it is "over the middle of the world" at the "middle of the day" embraces her and she becomes pregnant (Steven-

son 1894: 44–45). This theme of the impregnating power of the sun's rays is also evident in the tale of the Sun's son who, as a baby, identifies his father by crawling to the rays of the sun on the kiva floor (Alfonso Ortiz, personal communication 1981; Parsons 1940: 56–57).

The use of the rattlesnake as a marked glyph is consistent with the snake's wide ceremonial use and symbolic meaning in historical Pueblo culture, including a close association with the sun. A rattlesnake effigy and a ceremonial staff with the rattles of a rattlesnake indicated on it were found at Pueblo Bonito, which suggest a possible ritual involvement with the rattlesnake as part of the Chaco culture (Pepper 1920: 147). Among its many roles, the snake connects the below and above worlds (Tyler 1964: 222, 234). At one pueblo it is connected with the zenith and the sun: "Huwaka (Serpent of the Heavens) has a body like a crystal, and it is so brilliant that one's eyes cannot rest upon him; he is very closely allied to the sun" (Stevenson 1894: 69). Similarly, at another pueblo a snake with glistening scales is said to fly up to the sun each day and brilliantly reflect the sun's rays (Charles Loloma, a Hopi religious leader, personal communication 1983). Some reports on Pueblo cultures also connect the snake with fertility and equinox (Tyler 1964: 228–229, 245–247).

While there is no conclusive information concerning interpretations of the spiral among the historical Pueblos, some Pueblo people have indicated that it conveys the movement of people and clans. In one report, the spiral was described as the movement of people in their search for the center of the earth in the origin story (Roberts 1932: 151). The spiral has also been reported to represent the annual movement of the sun (Charles Loloma, personal communication 1983).

ASSOCIATIONS WITH THE PREHISTORIC PUEBLO CULTURE AND CHACO SOCIETY

The Fajada markings were probably developed when Pueblo people settled and flourished in the canyon between about A.D. 900 and 1300. The spiral petroglyph is identified with prehistoric

Pueblo people in this region during this period (Schaafsma 1980: 135–136; Polly Schaafsma, personal communication 1983).

The achievements of the Chaco culture — elaborate roads, irrigation works, and multistory architecture, all constructed between A.D. 1000 and 1150 — involved highly developed skills of planning, engineering, and surveying (Kincaid 1983; Lekson 1984; Marshall *et al.* 1979; Powers *et al.* 1983; Vivian 1974; also see chapter 1 in Carlson and Judge, *Astronomy and Ceremony*). During this period the Chaco society used these skills to align accurately several major constructions to the cardinal directions. Primary elements in the symmetric designs of the isolated great kiva at Casa Rinconada and of one of the central pueblos, Pueblo Bonito, and the initial portion of the North Road are oriented to within 0.25–0.5 degrees of the cardinal points (Sofaer *et al.* 1986; Stein 1983: 8–1; Williamson *et al.* 1977: 208–212; the canyon alignments were confirmed by surveys carried out by The Solstice Project).

The north/south alignments are particularly significant to the phenomenon of noon markings. A likely method of identifying noon is to watch shadows from a vertical object cross the north/south line as the sun crosses the meridian. Knowledge of true north within 0.25 degrees permits knowledge of solar noon at the latitude of Chaco to within less than one minute throughout the year. Similarly, knowledge of east and west to within less than 0.5 degrees allows the determination of the day of equinox. The architectural alignments indicate the Chaco culture's interest in and capability of achieving such accuracy. These alignments may also represent divisions of space that correspond to the markings on Fajada that divide the day and the year.

CONCLUSION

Although the astronomical markings and alignments of Chaco incorporate utilitarian calendric information, they do so with a redundancy and accuracy far beyond the practical requirements of time-keeping devices. For example, precise noon markings high on a steep butte serve no apparent useful purpose, nor do the markings of the lunar standstill cycle. The accurate alignments that define the major axes of the central ceremonial structures of the canyon are similarly abstract; for example, the east-west alignments could not have been used to determine (or have been determined by) equinox sunrise/sunset because of the locally elevated horizons. Similarly, the North Road was built elaborately and accurately in a direction that serves no apparent utilitarian purpose (Sofaer *et al.* 1986; Stein 1983: 8–1). Rather, what may be seen in the alignments and markings is the geometric expression of astronomical concepts and of the culture's cosmology.

Chaco Canyon was the center of an extensive road network and outlier system. Recent analysis indicates that it may have been a center for pilgrimage and ritual (Carlson and Judge, *Astronomy and Ceremony in the Prehistoric Southwest*, chapter 1). Fajada Butte may have been a center for the culture's ritual activity related to the sun and the moon. The clustering of markings found there is so far unique; surveys by the Solstice Project of all the prominent landforms in the Chaco cultural region have disclosed no further astronomical marking sites. When the sun in "the middle of the day" is over "the middle of the earth," the butte's glyphs commemorate this special moment in time and space. The extremes and midpoints of the solar and lunar cycles are integrated on Fajada. The daily and seasonal passages of the sun are united, as are the sun and moon and the earth and sky in the play of shadow and light on the rock carvings atop Fajada Butte.

ACKNOWLEDGMENTS

We are indebted again to Walter Herriman, former superintendent, and the staff of the Chaco Culture National Historic Park for their continued help and cooperation. Peggy Wier, Michael Marshall, Karl Kernberger, James Grant, Colin Franklin, Kenneth Butterfield, and Sheila Rotner helped collect the data, often under adverse conditions. We are grateful to Alfonso Ortiz, Gerald Hawkins,

Fred Eggan, Cesare Marino, Rob Blair, Evelyn Newman, Robert Mark, Harold Malde, Fred Nials, and particularly LeRoy Doggett and Stephen Mc-Cluskey for a number of helpful conversations. Pat Kenny is to be especially thanked for preparing the illustrations. We again thank our families and friends for their assistance and understanding.

[Note: We understand that an accompanying paper (published in Chapter 5 in Astronomy and Ceremony in the Prehistoric Southwest) *offers other explanations for some of the phenomena discussed here. We have not been given an opportunity to review this alternative analysis or to comment on it.]*

References Cited

Aveni, Anthony
    1980. *Skywatchers of Ancient Mexico.* University of Texas Press, Austin.

Benedict, Ruth Fulton
    1931. *Tales of the Cochiti Indians.* Bureau of American Ethnology Bulletin 98. Washington D.C.

Benedict, Ruth Fulton
    1935. *Zuñi Mythology.* Columbia University Contributions to Anthropology 21. New York.

Berry, Michael S.
    1982. *Time, Space and Transition in Anasazi Prehistory.* University of Utah Press, Salt Lake City.

Boas, Franz
    1925–1928. *Keresan Texts* (2 parts). *Publications of the American Ethnological Society* 8. New York.

Carlson, John B., and W. James Judge
    1983. *Astronomy and Ceremony in the Prehistoric Southwest.* Maxwell Museum of Anthropology, Anthropological Papers No. 2, University of New Mexico, Albuquerque.

Cordell, Linda S.
    1984. *Prehistory of the Southwest.* Academic Press, New York.

Curtis, Edward S.
    1926. *The North American Indian* 17. The Plimpton Press, Norwood, Mass.

Cushing, Frank M., compiler and translator
    1931. *Zuñi Folk Tales* (reprint). Alfred A. Knopf, New York. Originally published in 1901 by G. P. Putnam's Sons, New York.

Cushing, Frank M., compiler and translator
    1979. *Zuñi* (reprint), edited by Jesse Green. University of Nebraska Press, Lincoln. Originally published in 1882 by The Peripatetic Press, Santa Fe.

Dumarest, Noel
    1919. *Notes on Cochiti, New Mexico.* Memoirs of the American Anthropological Association 6(3), edited by Elsie C. Parsons. Lancaster, Pa.

Ellis, Florence H.
    1975. "A Thousand Years of the Pueblo Sun-Moon-Star Calendar," in *Archaeoastronomy in Precolumbian America,* edited by A. F. Aveni, pp. 58–87. University of Texas Press, Austin.

Gunn, John M.
    1917. *Schat-chen: History, Traditions and Narratives of the Queres Indians of Laguna and Acoma.* Albright and Anderson, Albuquerque.

Hayden, F. V., editor
    1878. *Tenth Annual Report—Geological and Geographical Survey of the Territories.* Washington, D. C.

Hewett, Edgar L.
    1936. *The Chaco Canyon and Its Monuments.* University of New Mexico Press, Albuquerque.

Kallestewa, B., J. Niiha, A. Peywa, A. Pinto, R. Quam, A. Nastacio, and W. Eriacho
    1984. Statements by Zuñi religious leaders on Kolhu wala:wa. Testimony presented to the Select Committee on Indian Affairs, U.S. Senate, April 3, 1984. Washington, D. C.

Kincaid, Chris, editor
    1983. *Chaco Roads Project Phase I.* Bureau of Land Management, Albuquerque.

Krupp, E. C.
    1977. *In Search of Ancient Astronomies.* Doubleday, Garden City, New York.

Krupp, E. C.
    1983. *Echoes of the Ancient Skies.* Harper and Row, New York.

Lange, Charles M.
    1959. *Cochiti: A New Mexico Pueblo, Past and Present.* University of Texas Press, Austin.

Lekson, Stephen
    1984. *Great Pueblo Architecture of Chaco Canyon, New Mexico.* National Park Service Publications in Archeology 18B. Albuquerque.

Lowie, R. H.
    n.d. "Hopi Indian Inside a Kiva." Photo No. 283545. American Museum of Natural History, New York.

Marshall, Michael P., John Stein, Richard Loose, and Judith E. Novotny
1979. *Anasazi Communities of the San Juan Basin.* Public Service Company of New Mexico, Albuquerque, and New Mexico State Historic Preservation Bureau, Santa Fe.

Mindeleff, Victor
1891. "A Study of Pueblo Architecture," in *Eighth Annual Report of the Bureau of American Ethnology, 1886–1887* pp. 3–228. Washington, D. C.

Newcomb, F. J.
1966. *Navajo Neighbors.* University of Oklahoma Press, Norman.

Newman, E. B., R. K. Mark, and R. G. Vivian
1982. "Anasazi Solar Marker: The Use of a Natural Rockfall," *Science* 217: 1036–1038.

Ortiz, Alfonso
1969. *The Tewa World: Space, Time, Being, and Becoming in a Pueblo Society.* University of Chicago Press, Chicago.

Parmentier, Richard J.
1979. "The Pueblo Mythological Triangle: Poseyemu, Montezuma, and Jesus in the Pueblos," in *Southwest,* edited by Alfonso Ortiz, pp. 609–622. *Handbook of North American Indians,* Vol. 9. Smithsonian Institution, Washington, D. C.

Parsons, Elsie Clews
1929. *The Social Organization of the Tewa of New Mexico.* Memoirs of the American Anthropological Association 36. Menasha, Wis.

Parsons, Elsie Clews
1932. "Isleta, New Mexico," in *Forty-Seventh Annual Report of the Bureau of American Ethnology for the Years 1929–1930,* pp. 193–466. Washington, D. C.

Parsons, Elsie Clews
1940. *Taos Tales.* Memoirs of the American Folklore Society 34. G. F. Stechert, New York.

Pepper, George M.
1920. *Pueblo Bonito.* Anthropological Papers of the American Museum of Natural History 27. New York.

Powers, R. P., W. B. Gillespie, and S. H. Lekson
1983. *The Outlier Survey.* Reports of the Chaco Center 3. National Park Service, Division of Cultural Research, Albuquerque.

Reyman, Jonathan E.
1976. "Astronomy, Architecture, and Adaptation at Pueblo Bonito," *Science* 193: 957–962.

Roberts, F. H. H., Jr.
1932. *The Village of the Great Kivas on the Zuñi Reservation, New Mexico.* Bureau of American Ethnology Bulletin 111. Washington, D. C.

Schaafsma, Polly
1980. *Indian Rock Art of the Southwest.* University of New Mexico Press, Albuquerque.

Sofaer, A., and J. Crotty
1977. *Survey of Rock Art on Fajada Butte.* Manuscript on file, Chaco Center, University of New Mexico, Albuquerque.

Sofaer, A., M. P. Marshall, and R. M. Sinclair
1986. *Cosmographic Expression in the Road System of the Chaco Culture of Northwestern New Mexico.* Paper presented at the Second Oxford Conference on Archaeoastronomy, Merida (Yucatan), Mexico.

Sofaer, A., R. M. Sinclair, and L. E. Doggett
1982a. "Lunar Markings on Fajada Butte," in *Archaeoastronomy in the New World,* edited by A. Aveni, pp. 169–181. Cambridge University Press, Cambridge and New York.

Sofaer, A., R. M. Sinclair, and L. E. Doggett
1982b. "Noon Markings on Fajada Butte, Chaco Canyon, New Mexico," *Bulletin of the American Astronomical Society* 14:872.

Sofaer, A., V. Zinser, and R. M. Sinclair
1979. "A Unique Solar Marking Construct," *Science* 206: 283–291.

Stein, John R.
1983. "Road Corridor Descriptions," in *Chaco Roads Project Phase I,* edited by C. Kincaid, pp. 8-1 to 8-15. Bureau of Land Management, Albuquerque.

Stephen, A. M.
1929. "Hopi Tales," *Journal of American Folklore* 42: 1–72.

Stevenson, Matilda Coxe
1894. "The Sia," in *Eleventh Annual Report of the Bureau of American Ethnology for the Years 1889–1890,* pp. 3–157. Washington, D. C.

Stevenson, Matilda Coxe
1904. "The Zuñi Indians: Their Mythology, Esoteric Fraternities, and Ceremonies," in *Twenty-Third Annual Report of the Bureau of American Ethnology for the Years 1901–1902,* pp. 3–634. Washington, D. C.

Tedlock, Barbara
1984. "Zuñi Sacred Theater," *American Indian Quarterly* 7(1): 93–110.

Titiev, Mischa
1944. *Old Oraibi: A Study of the Hopi Indians of Third Mesa.* Papers of the Peabody Museum of American Archaeology and Ethnology 22(1). Harvard University, Cambridge, Mass.

Tyler, Hamilton A.
1964. *Pueblo Gods and Myths.* University of Oklahoma Press, Norman.

Upham, Steadman
1984. *Politics and Power—An Economic and Political History of the Western Pueblo.* Academic Press, New York.

Vivian, R. Gwinn
1974. "Conservation and Diversion: Water-Control Systems in the Anasazi Southwest," in *Irrigation's Impact on Society,* edited by Theodore Dawning and McGuire Gibson, pp. 95–112. Anthropological Papers of the University of Arizona 25. Tucson.

Williamson, Ray A.
1981. "North America: A Multiplicity of Astronomies," in *Archaeoastronomy in the Americas,* edited by R. A. Williamson, pp. 61–80. Ballena Press, Los Altos, Calif.

Williamson, R. A., H. J. Fisher, and D. O'Flynn
1977. "Anasazi Solar Observatories," in *Native American Astronomy,* edited by A. F. Aveni, pp. 203–217. University of Texas Press, Austin.

4

Pueblo Bonito Petroglyph on Fajada Butte: Solar Aspects

Summary: *Astronomical use of petroglyphs on Fajada Butte, in Chaco Canyon, New Mexico, is one of numerous expressions of solar-lunar cosmology by the Chacoan culture. Earlier work showed that 13 light markings on petroglyphs record the extremes and mid-positions of the solar and lunar cycles. Recent work has shown that there are also redundant expressions of this solar-lunar astronomy in the orientations, internal geometry, and inter-building relationships of the 14 major Chacoan buildings. The small and intimate scale of the display of light markings on Fajada Butte contrasts with the massive and public scale of the Chacoans' expression of astronomy in their buildings. And yet, a small petroglyph near the top of the Fajada Butte, which conveys the symbolic relationship of the sun to Pueblo Bonito, provides insight to the design and purpose of the most massive construction of the Chacoan culture.*

This paper was presented at the 1994 International Rock Art Congress ("Rock Art-World Heritage"), held in Flagstaff, Arizona.

A PETROGLYPH NEAR THE TOP OF FAJADA BUTTE in Chaco Canyon, New Mexico, appears to display the major features of the ground plan of Pueblo Bonito and to express the interest of the prehistoric Chacoan culture in the solar annual and diurnal cycle (Sofaer and Sinclair 1989) (Figures la, 1b; 2).

In this regard, we consider three aspects of the Pueblo Bonito petroglyph: its similarity to the ground plan of Pueblo Bonito, the petroglyph's own solar orientations, and its association with the motif of other solar petroglyphs near the top of Fajada Butte.

Our interpretation of the petroglyph is consistent with, but does not depend on, cultural evidence of the association of its arrow (and possibly bow-and-arrow) image with the solar cosmology of the descendant Pueblo peoples during the historic period.

Anna Sofaer

Published in Celestial Seasonings: Astronomical Connotations of Rock Art, *a section in* International Rock Art Congress Proceedings 1994, *E. C. Krupp, section editor (American Rock Art Research Association, 2006)*

FIGURE 1A. *This petroglyph near the top of Fajada Butte is 24 cm by 36 cm and is located about 10 m west of the three-slab Sun Dagger site (see Sofaer et al. 1979). A, B, C, & D indicate features of Pueblo Bonito, shown in the ground plan in* FIGURE 1B. *Note: The petroglyph is artificially emphasized because of its inherent low contrast.*

BACKGROUND

During the period A.D. 900-1130, the Chacoan culture built numerous multistoried buildings and extensive roads throughout the 80,000 square kilometers of the San Juan Basin of northwestern New Mexico (Lekson *et al.* 1988). Chaco Canyon was the center of the culture. Recent archaeological interpretations suggest that several of the large central buildings, including Pueblo Bonito, were used primarily for ceremony and that Chaco Canyon served as a ceremonial center for outlying Chacoan communities (Judge 1984).

Astronomy played an important role in the Chacoan culture. Earlier research showed that the Chacoans marked the solstices, the equinoxes, the sun's local solar noon, and the standstill positions of the moon in 13 light markings on petroglyphs on Fajada Butte (Sofaer *et al.* 1979, 1982; Sofaer and Sinclair 1987). Recent studies show that 12 of the

major Chacoan buildings are oriented to the sun and moon (Sofaer *et al.* 1998) and that each of the major buildings also has an internal geometry that corresponds to the relationships of the solar and lunar cycles (Sofaer 1994). In addition, most of the major buildings are organized in a solar and lunar pattern that is ordered about Chaco Canyon (Sofaer 1998). Pueblo Bonito, which is located at the approximate center of Chaco Canyon, plays a central role in this pattern.

Pueblo Bonito was originally four stories high and contained more than 700 rooms (Lekson 1984). It also contained 36 kivas, the round semi-subterranean features of the Chacoan culture, which are thought to have served ceremonial functions. In its longest dimension, Pueblo Bonito is 150 meters. It was the most massive structure in the Chacoan cultural region. Its semicircular shape is unique among Chacoan buildings.

PUEBLO BONITO AND ITS PETROGLYPHIC LIKENESS

The primary external design of Pueblo Bonito is a near-semicircle composed of a long front wall, or diameter of the semicircle, which is joined at each end by a long curved back wall, or arc of the semicircle.

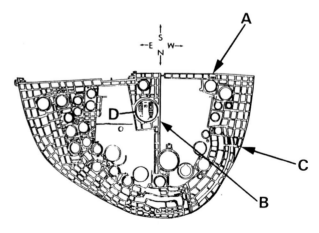

FIGURE 1B. *In this ground plan of Pueblo Bonito key features are coded by letter. A is the east-west oriented wall. B is the north-south oriented wall. C is the semicircular outline of the building. D is Kiva A. (Copyright © 1995 Solstice Project.)*

FIGURE 2. *An aerial view of Pueblo Bonito from the north permits a comparison to the characteristics of the petroglyph.* *(Copyright © David L. Brill.)*

The primary internal feature consists of a wall which is perpendicular to the diameter of the semicircle and which approximately bisects it if the line of the interior wall is extended to the long curved back wall as a radius of the semicircle. In addition, Kiva A, a prominent internal feature, is located near the juncture of the primary internal, or radial, wall and the long front, or diameter, wall. Kiva A is 17 m in diameter. It is the largest exposed kiva in Pueblo Bonito.[1]

Fajada Butte, a 135-meter high promontory, is located at the south entrance of Chaco Canyon, seven kilometers to the southeast of Pueblo Bonito.

The Pueblo Bonito petroglyph is located at the approximate center of the highest cliff bench of the Butte, about 10 m to the west and slightly above the three-slab Sun Dagger site. Pecked onto a vertical cliff surface, its center is 114 centimeters from the present floor of the cliff bench. The petroglyph is 24 cm by 36 cm.

The petroglyph appears to map the major features of the ground plan of Pueblo Bonito. The design is a near-semicircle. Its diameter corresponds to the long front wall and its arc corresponds to the long curved back wall of Pueblo Bonito. This semicircular shape is unique among the petroglyphs on Fajada Butte. The uniqueness of the semicircular format of the ground plan and of the petroglyph calls further attention to their relationship.

The primary internal features of the petroglyph consist of a radial line perpendicular to the diameter of the semicircle, approximately bisecting it, and a drill hole, approximately 1.7 cm deep, near the juncture of the diameter and the radial line perpendicular to it. The radial line appears to correspond to the interior wall and its extension. The drill hole appears to correspond to Kiva A. (We note with interest that the east and west sides of Pueblo Bonito are represented in the petroglyph in reverse of their positions in the ground plan. Kiva A, however, is represented in the petroglyph in its position in the ground plan.[2])

Two features of the petroglyph are external to the semicircle. First, the head and butt of an arrow are appended to the radial line of the semicircle. In this context, the radial line can be seen as the shaft of the arrow; and the semicircle can possibly be seen as a bow. (We note that, if the image is interpreted as a bow-and-arrow, the order of the bow and arrow are the reverse of what we find when a bow and arrow are more ready for use.) Second, a spiral surmounts the diameter of the semicircle.

SOLAR ASPECTS OF THE PETROGLYPH

We identify several lines of evidence for solar aspects of the Pueblo Bonito petroglyph.

Similarity to Pueblo Bonito. The broad similarity of the petroglyph to the ground plan of Pueblo Bonito—the petroglyph's semicircular shape, the radial line dividing it, and the drill hole corresponding to Kiva A—recalls the solar orientation of Pueblo Bonito itself and of Kiva A.

Earlier work shows that the western segment of Pueblo Bonito's long front wall (or diameter wall) is oriented east-west, the directions of the rising and setting sun at equinox (Sofaer 1998). This work also shows that the primary internal feature—the radial wall which is perpendicular to the western segment of the diameter wall—is

oriented north-south and to the azimuth of the meridian passage of the sun at noon. The internal features of Kiva A are also oriented to the cardinal directions. It may be further considered that the vertical axis of Kiva A relates to the nadir and zenith directions.

As a representation of the key solar elements of Pueblo Bonito's design the petroglyph can be seen to express solar relationship. It refers to the mid-point in the sun's yearly passage and to the mid-point in the sun's daily passage.[3] It can also be seen, in representing Kiva A, as expressing the nadir and the zenith.

The Petroglyph's Orientations. The Pueblo Bonito petroglyph is oriented to the cardinal directions and to the zenith and nadir. The petroglyph faces east to the rising sun at equinox, and the diameter of its semicircle is oriented north-south to the azimuth of the sun's meridian passage. Its radial line is vertical and thus is oriented to the zenith and the nadir. The drill hole—which appears to be associated with Kiva A—penetrates, in a westward direction, the cliff surface as Kiva A penetrates the floor of the canyon.

Each of the cardinal orientations of the building is transformed in the petroglyph to another cardinal direction. The east-west diameter wall of Pueblo Bonito is north-south in the petroglyph. The north-south interior wall of Pueblo Bonito is the nadir-zenith in the petroglyph. The nadir-zenith orientation of Kiva A is east-west in the petroglyph. Each day, over a period of about a half-hour during meridian passage of the sun, the visibility of the petroglyph changes gradually from full sunlight to full shadow. This light and shadow transformation does not involve a distinct marking on the petroglyph. We are, therefore, not certain of its significance to the Chacoans. It seems likely, however, that the effect was noticed by the Chacoans who used numerous light and shadow interactions to record the sun's meridian passage on petroglyphs on Fajada Butte. It may in fact have been an effect intended by the Chacoans in their placement of the petroglyph on a cliff face with a north-south orientation.

It is of interest that at summer solstice during the sun's meridian passage, a stick inserted in the drill hole of the petroglyph casts a vertical shadow that falls approximately along the radial line.[4] (Such an inserted stick would, itself, point approximately 90 degrees away from the astronomical zenith to the equinoctial sunrise.) Again, we note this with interest, but cannot be certain of its significance to the Chacoans.

Associated Features. The spiral which curls above the diameter line of the petroglyph and to which the arrow points may represent an additional solar feature. We note that the association of spirals with solar events is exhibited repeatedly on Fajada Butte. Eleven markings of shadow and light on petroglyphs near the top of Fajada Butte commemorate the solstices, equinoxes, and meridian passage of the sun. Eight of these markings occur on spiral petroglyphs.

We note that in the petroglyph the arrow points to the spiral, which seems to have solar associations, and to the zenith.[5]

CULTURAL SIMILARITIES IN THE HISTORIC PERIOD

The arrow and the bow-and-arrow are associated with the sun in the cosmology of the historic Pueblo peoples, who are descendants of the Chacoan people. In certain Pueblo traditions, the arrow is seen as a vertical axis that relates to the nadir and the zenith... or to the world below and above. In a version of the Zuñi creation story, the Sun-Father gives his sons bows and arrows and directs them to lift, with an arrow, the Sky-Father to the zenith (Cushing 1896). In other versions, the sun directs his sons to use their bows and arrows to open the way to the world below for the Pueblo people to emerge to the earth's surface and to the sun's light (Bunzel 1932). At the solstices, the Pueblo people give offerings of miniature bows and arrows to the sun (Ellis and Hammack 1968; Parsons 1939). The sun is depicted by Pueblo groups as carrying a bow and arrow (Stevenson 1894).

SUMMARY

A number of features of the Pueblo Bonito petroglyph, taken together, powerfully suggest its solar character.[6]

ACKNOWLEDGMENTS

We appreciate the consultations and assistance in field observations of Rolf M. Sinclair (NSF) and Dabney Ford (NPS). We were greatly assisted by an astute editor, known as A. Sloan. We thank the late Alfonso Ortiz, the late Fred Eggan, John Stein, and Stephen Lekson for many helpful discussions. This paper and others from the Solstice Project can be read at *www.solstice project.org.*

NOTES

1. Kiva K, which was covered by the Chacoans in the late 1000s, is the only kiva in Pueblo Bonito that was larger than Kiva A.

2. Another possible explanation for the position of the drill hole on the petroglyph is that it may represent Kiva K. This Kiva was located on the side opposite to Kiva A of the radial wall, also near the radial wall's juncture with the diameter wall.

3. See John Stein, in *Anasazi Architecture and American Design,* University of New Mexico Press, 1998, for another view on the importance of solar noon in the design and location of Pueblo Bonito.

4. We note that Dabney Ford (National Park Service) has also observed this phenomenon.

5. In our consideration of the solar aspects of the petroglyph, it is of particular interest that the interior wall of Pueblo Bonito, which is oriented to the azimuth of the meridian passage of the sun, is represented in the petroglyph as an arrow pointing to the zenith. This possible association of the zenith-pointing arrow with the meridian passage of the sun also suggests a link between the nearby Sun Dagger (which occurs 35 minutes before solar noon at summer solstice) and the seven markings of solar noon on Fajada Butte with Pueblo Bonito. Ultimately, this petroglyph may deepen our understanding of the design of Pueblo Bonito.

6. We make special note of the rotations and reflections in the petroglyph. Our future study of this petroglyph will focus on these interesting characteristics.

REFERENCES CITED

Bunzel, R. L.
1932. *Zuñi Origin Myths.* Forty-seventh Annual Report Bureau of American Ethnology, p. 584. Smithsonian Institution, Washington, D.C.

Cushing, F. H.
1896. *Outlines of Zuñi Creation Myths.* Thirteenth Annual Report, Bureau of American Ethnology, p. 382. Smithsonian Institution, Washington, D.C.

Ellis, F. H. and L. Hammack
1968. "The Inner Sanctum of Feather Cave," *American Antiquity* 331: 25, 44.

Judge, Jr., W. J.
1984. "New Light on Chaco," in *New Light on Chaco,* edited by D. G. Noble, pp. 1-12. School of American Research, Santa Fe.

Lekson, S. H.
1984. *Great House Architecture of the Chacoan Culture.* National Park Service, Albuquerque.

Lekson, S. H. , T. C. Windes, J. R. Stein, and
       W. J. Judge, Jr.
1988. "The Chaco Canyon Community," *Scientific American,* July, pp. 100-109.

Parsons, E. C.
1939. *Pueblo Indian Religion.* University of Chicago Press, Chicago.

Sofaer, A.
1994. "Chacoan Architecture: A Solar-Lunar Geometry," in *Time and Astronomy at the Meeting of Two Worlds,* edited by S. Iwaniszewski *et al.,* pp. 265-278. Warsaw University, Center for Latin American Studies, Warsaw, Poland.

Sofaer, A.
1998. "The Primary Architecture of the Chacoan Culture: A Cosmological Expression," in *Anasazi Architecture and American Design,* edited by B. Morrow and V. B. Price. University of New Mexico Press, Albuquerque.

Sofaer, A., and R. M. Sinclair
1987. "Astronomical Markings at Three Sites on Fajada Butte, Chaco Canyon, New Mexico," in *Astronomy and Ceremony in the Prehistoric Southwest,* edited by J. Carlson and W. J. Judge, Jr., pp. 13-70. Maxwell Museum of Anthropology, Albuquerque.

Sofaer, A., and R. M. Sinclair
1989. "An Interpretation of a Unique Petroglyph in Chaco Canyon, New Mexico," in *World Archaeoastronomy,* edited by A F. Aveni, p. 499. Cambridge University Press, Cambridge, England.

Sofaer, A., R. M. Sinclair, and A. Doggett
1982. "Lunar Markings on Fajada Butte, Chaco Canyon, New Mexico," in *Archaeoastronomy in the New World,* edited by A. F. Aveni, pp. 169-181. Cambridge University Press, Cambridge, England.

Sofaer, A., V. Zinser, and R. M. Sinclair
1979. "A Unique Solar Marking Construct," *Science* 206: 283-291.

Stevenson, M. C.
1894. *The Sia.* Eleventh Annual Report Bureau of American Ethnology, p. 35, Smithsonian Institution, Washington, D.C.

Astronomical Expressions in the Major Chacoan Buildings

5

The Primary Architecture of the Chacoan Culture:

A Cosmological Expression

"The Primary Architecture of the Chacoan Culture: A Cosmological Expression" most recently appeared as a chapter in The Architecture of Chaco Canyon, New Mexico, *Stephen H. Lekson, editor (University of Utah Press, Salt Lake City, 2007). It was published earlier in* Anasazi Architecture and American Design, *Baker Morrow and V. B. Price, editors (University of New Mexico Press, Albuquerque, 1997).*

Anna Sofaer

STUDIES BY THE SOLSTICE PROJECT indicate that the major buildings of the ancient Chacoan culture of New Mexico contain solar and lunar cosmology in three separate articulations: their orientations, internal geometry, and geographic interrelationships were developed in relationship to the cycles of the sun and moon.

From approximately 900 to 1130, the Chacoan society, a prehistoric Pueblo culture, constructed numerous multistoried buildings and extensive roads throughout the eighty thousand square kilometers of the arid San Juan Basin of northwestern New Mexico (Cordell 1984; Lekson *et al.* 1988; Marshall *et al.* 1979; Vivian 1990) (Figure 9.1). Evidence suggests that expressions of the Chacoan culture extended over a region two to four times the size of the San Juan Basin (Fowler and Stein 1992; Lekson *et al.* 1988). Chaco Canyon, where most of the largest buildings were constructed, was the center of the culture (Figures 9.2 and 9.3). The canyon is located close to the center of the high desert of the San Juan Basin.

Twelve of the fourteen major Chacoan buildings are oriented to the midpoints and extremes of the solar and lunar cycles (Sofaer, Sinclair, and Donahue 1991). The eleven rectangular major Chacoan buildings have internal geometry that corresponds to the relationship of the solar and lunar cycles (Sofaer, Sinclair, and Donahue 1991). Most of the major buildings also appear to be organized in a solar-and-lunar regional pattern that is symmetrically ordered about Chaco Canyon's central complex of large ceremonial buildings (Sofaer, Sinclair, and Williams 1987). These findings suggest a cosmological purpose motivating and

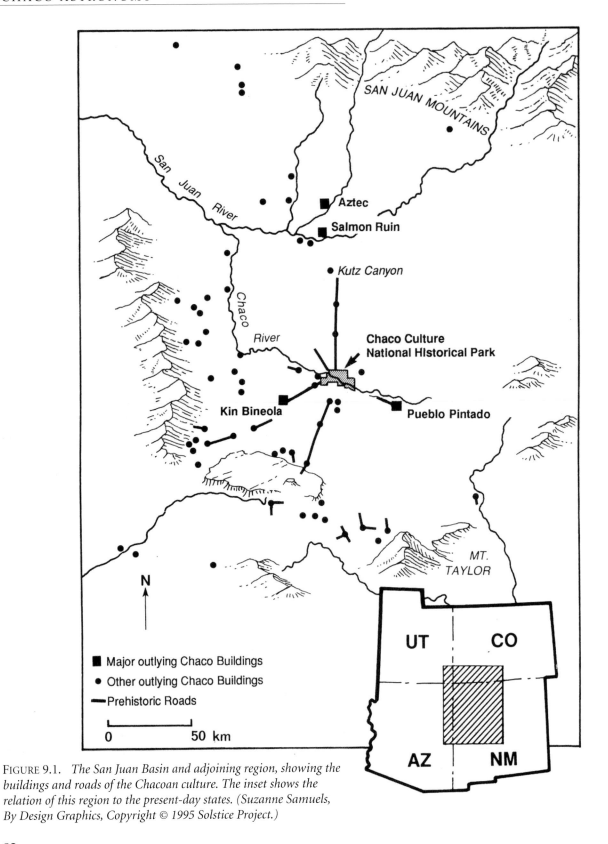

FIGURE 9.1. *The San Juan Basin and adjoining region, showing the buildings and roads of the Chacoan culture. The inset shows the relation of this region to the present-day states. (Suzanne Samuels, By Design Graphics, Copyright © 1995 Solstice Project.)*

directing the construction and the orientation, internal geometry, and interrelationships of the primary Chacoan architecture.

This essay presents a synthesis of the results of several studies by the Solstice Project between 1984 and 1997 and hypotheses about the conceptual and symbolic meaning of the Chacoan astronomical achievements. For certain details of Solstice Project studies, the reader is referred to several earlier published papers.[1]

BACKGROUND

The Chacoan buildings were of a huge scale and "spectacular appearance" (Neitzel 1989). The buildings typically had large public plazas and elaborate "architectural earthworks" that formed road entries (Stein and McKenna 1988). The major Chacoan buildings, the subject of the Solstice Project's studies (Figures 9.3 and 9.4), are noted in particular for their massive core veneer masonry. They were up to four stories high and contained as many as seven hundred rooms, as well as numerous kivas, including great kivas, the large ceremonial chambers of prehistoric Pueblo culture (Lekson 1984; Marshall *et al.* 1979; Powers, Gillespie, and Lekson 1983).

The construction of the major Chacoan buildings employed enormous quantities of stone and wood. For example, 215,000 timbers—transported from distances of more than 80 kilometers—were used in the canyon in the major buildings alone (Lekson *et al.* 1988). The orderly, gridlike layout of the buildings suggests that extensive planning and engineering were involved in their construction (Lekson 1984; Lekson *et al.* 1988).

No clear topographic or utilitarian explanations have been developed for the orientations of the Chacoan buildings. The buildings stand free of the cliffs, and their specific orientations are not significantly constrained by local topography.[2] Although the need to optimize solar heating may have influenced the general orientations of the buildings, it probably did not restrict their orientations to specific azimuths. Similarly, environmental factors, such as access to water, appear not to have

dominated or constrained the Chacoans' choice of specific locations for their buildings.[3]

The Chacoans also constructed more than two hundred kilometers of roads. The roads were of great width (averaging 9 meters wide), and they were developed, with unusual linearity, over distances of up to fifty kilometers. Their construction required extensive surveying and engineering (Kincaid 1983). Investigations show that certain of the roads were clearly overbuilt if they were intended to serve purely utilitarian purposes (Lekson 1991; Roney 1992; Sofaer, Marshall, and Sinclair 1989; Stein 1989),[4] and that they may have been constructed as cosmographic expressions (Marshall 1997; Sofaer, Marshall, and Sinclair 1989).

Scholars have puzzled for decades over why the Chacoan culture flourished in the center of the desolate environment of the San Juan Basin. Earlier models proposed that Chaco Canyon was a political and economic center where the Chacoans administered a widespread trade and redistribution system (Judge 1989; Sebastian 1992). Recent archaeological investigations show that major buildings in Chaco Canyon were not built or used primarily for household occupation (Lekson *et al.* 1988). This evidence, along with the dearth of burials found in the canyon, suggests that, even at the peak of the Chacoan development, there was a low resident population. (Estimates of this population range from 1,500 to 2,700 [Lekson 1991; Windes 1987].) Evidence of periodic, large-scale breakage of vessels at key central buildings indicates, however, that Chaco Canyon may have served as a center for seasonal ceremonial visitations by great numbers of residents of the outlying communities (Judge 1984; Toll 1991).

Many aspects of the Chacoan culture—such as the transport of thousands of beams and pots—have struck archaeologists as having a "decided aura of inefficiency" (Toll 1991). Other findings—such as "intentionally destroyed items in the trash mounds," "plastered-over exquisite masonry," and strings of beads "sealed into niches" in a central great kiva—indicate esoteric uses of Chacoan constructions. It has been suggested that, in the "absence of any evidence that there is either a

FIGURE 9.2. *Aerial view of central area of Chaco Canyon, looking north. The photograph shows three major buildings: Pueblo Bonito (left), Pueblo Alto (above center), and Chetro Ketl (right). Casa Rinconada (bottom center) and New Alto are also shown. (Photograph by Adriel Heisey, Copyright © 1995 Adriel Heisey.)*

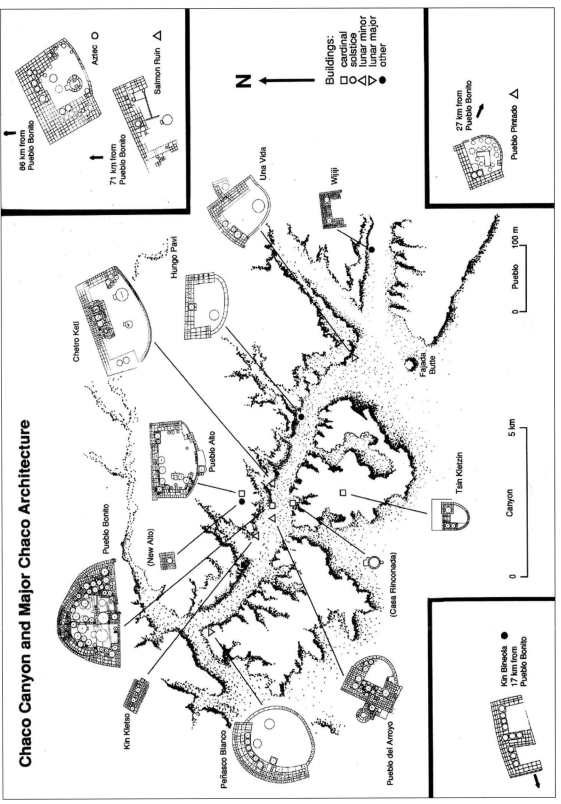

FIGURE 9.3. *Chaco Canyon, showing the locations and ground plans of ten major buildings (and two minor buildings). Four outlying major buildings are also shown. The astronomical orientations of the buildings are indicated. (Fabian Schmid, Davis, Inc.; and Suzanne Samuels, By Design Graphics, Copyright © 1995 Solstice Project.)*

FIGURE 9.4. *Aerial photographs of two major buildings in Chaco Canyon, Pueblo Bonito (upper) and Pueblo del Arroyo (lower) with ground plans of these buildings. (Photographs by Koogle and Pouls for the National Park Service; graphics by Suzanne Samuels, By D esign Graphics, Copyright © 1993 Solstice Project.)*

natural or societal resource to which Chaco could control access by virtue of its location" (Toll 1991), Chaco Canyon was the center of exchange of information and knowledge (Sebastian 1991). Two other archaeologists suggest that Chaco Canyon was a "central archive for esoteric knowledge, such as maintenance of the region's ceremonial calendar" (Crown and Judge 1991).

Scholars have commented extensively on the impractical and enigmatic aspects of Chacoan buildings, describing them as "overbuilt and overembellished" and proposing that they were built primarily for public image and ritual expression (Lekson *et al.* 1988; Stein and Lekson 1992). Some observers have thought that the Chacoan buildings were developed as expressions of the Chacoans' "concepts of the cosmos" (Stein and Lekson 1992) and that their placement and design may have been determined in part by "Chacoan cosmography" (Marshall and Doyel 1981). One report proposes that "Chaco and its hinterland are related by a canon of shared design concepts" and that the Chacoan architecture is a "common ideational bond" across a "broad geographic space" (Stein and Lekson 1992). That report suggests that the architectural characteristics of Pueblo Bonito, one of the two largest and most central buildings of the Chacoan system, are rigorously repeated throughout the Chaco region. Thus, important clues to the symbology and ideology of the Chacoan culture may be embedded in its central and primary architecture and expressed in the relationship of this architecture to primary buildings in the outlying region.

Numerous parallels to the Chacoan expressions of cosmology appear in the astronomically and geometrically ordered constructions of Mesoamerica—a region with which the Chacoans are known to have had cultural associations (Aveni 1980; Broda 1993). Moreover, traditions of the descendants of the prehistoric Pueblo people, who live today in New Mexico and Arizona, also suggest parallels to the Chacoan cosmology and give us insight into the general cosmological concepts of the Chacoan culture.

PREVIOUS WORK

Solstice Project studies, begun in 1978, documented astronomical markings at three petroglyph sites on Fajada Butte, a natural promontory at the south entrance of Chaco Canyon (Figure 9.3). Near the top of the butte, three rock slabs collimate light so that markings of shadow and light on two spiral petroglyphs indicate the summer and winter solstices, the equinoxes, and the extreme positions of the moon, that is, the lunar major and minor standstills (Sofaer, Zinser, and Sinclair 1979; Sofaer, Sinclair, and Doggett 1982; Sinclair *et al.* 1987). At two other sites on the butte, shadow and light patterns on five petroglyphs indicate solar noon and the solstices and equinoxes (Sofaer and Sinclair 1987).

A 1989 Solstice Project study showed astronomical significance in the Chacoans' construction of the Great North Road (Sofaer, Marshall, and Sinclair 1989). This nine-meters-wide, engineered road extends from Chaco Canyon north 50 kilometers to a badlands site, Kutz Canyon (Figure 9.1). The purpose of the road appears to have been to articulate the north-south axis and to connect the canyon's central ceremonial complex with distinctive topographic features in the north.

Prior to the Solstice Project studies of the Chacoan constructions, others had reported cardinal orientations in the primary walls and the great kiva of Pueblo Bonito, a major building located in the central complex of Chaco Canyon, and in Casa Rinconada, an isolated great kiva (Williamson *et al.* 1975, 1977). Researchers have also shown that certain features in Pueblo Bonito and Casa Rinconada may be oriented to the solstices (Reyman 1976; Williamson *et al.* 1977; Zeilik 1984).[5]

Certain early research also highlighted astronomically related geometry and symmetry in the Chacoan architecture. One scholar describes "geometrical/astronomical patterns" in the extensive cardinal organization of Casa Rinconada (Williamson 1984). His report notes that these patterns were derived from the symmetry of the solar cycle, rather than from the observation of astronomical events from this building. Similarly, other research describes a symmetric, cardinal patterning in the geographic relationships of several central buildings, and it further suggests that other major buildings—outside of the center and out of sight of the center—were organized in symmetric relationships to the cardinal axes of the center (Fritz 1978).

These previous findings led the Solstice Project to examine and analyze the orientations, internal geometry, and interrelationships of the major Chacoan buildings for possible astronomical significance. The Solstice Project's study regarded as important both orientations to visible astronomical events and expressions of astronomically related geometry. In the following analysis, the Solstice Project considers the orientations of the major Chacoan buildings, and of their interbuilding relationships, to astronomical events on both the sensible and the visible horizons.[6]

SOLAR AND LUNAR ORIENTATIONS OF THE MAJOR CHACOAN BUILDINGS

The Solstice Project asked if the fourteen major buildings were oriented to the sun and moon at the extremes and mid-positions of their cycles—in other words, the meridian passage, the solstices and the equinoxes, and the lunar major and minor standstills. The rising and setting azimuths for these astronomical events at the latitude of Chaco Canyon are given in Figure 9.5. (The angles of the solstices, equinoxes, and lunar standstills are expressed as single values taken east and west of north as positive to the east of north and negative to the west of north.)

In the clear skies of the high desert environment of the San Juan Basin, the Chacoans had nearly continuous opportunity to view the sun and the moon, to observe the progression of their cycles, and to see the changes in their relationships to the surrounding landscape and in patterns of shadow and light.

The sun: The yearly cycle of the sun is evident by its excursions to the extreme positions: rising in the northeast at the summer solstice and in the southeast at the winter solstice; setting in the northwest at the summer solstice and in the southwest at the winter solstice. At equinox, in the middle of these excursions, it rises and sets east and west. At solar noon, in the middle of its daily excursion, the sun is on the meridian—in other words, aligned with the north-south axis.

The cardinal directions (0°, 90°) are regarded in this paper as having the solar associations of equinox and meridian passage.[7] In a location surrounded by significantly elevated topography, however, the equinox sun can also be observed on the visible horizon in sunrise and sunset azimuths that are not the cardinal east-west axis of the sensible horizon.

The moon: The moon's standstill cycle is longer (18.6 years) and more complex than the sun's cycle, but its rhythms and patterns also can be observed in its shifting positions on the horizon, as well as in its relationship to the sun (see also Aveni 1980: Chapter 3). In its excursions each month it shifts from rising roughly in the northeast to rising roughly in the southeast and from setting roughly in the northwest to setting roughly in the southwest, but a closer look reveals that the envelope of these excursions expands and contracts through the 18.6-year standstill cycle. In the year of the major standstill, this envelope is at its maximum width, and at the latitude of Chaco, the moon rises and sets approximately 6.1° north and south of the positions of the rising and setting solstice suns. These positions are the farthest to the northeast and northwest and southeast and southwest that the moon ever reaches. In the year of the minor standstill, nine to ten years later, the envelope is at its minimum width, and the moon rises and sets approximately 6.7° within the envelope of the rising and setting solstice suns.

The progression of the sun and the moon in their cycles can also be quite accurately observed in their changing heights at meridian passage and in the accompanying shifts in shadow patterns.

A number of factors, such as parallax and atmospheric refraction, can shift and broaden the range of azimuth where the risings and settings of the solstice suns and the standstill moons appear on the horizon. In addition, judgments in determining a solar or lunar event introduce uncertainties. These judgments involve determining which

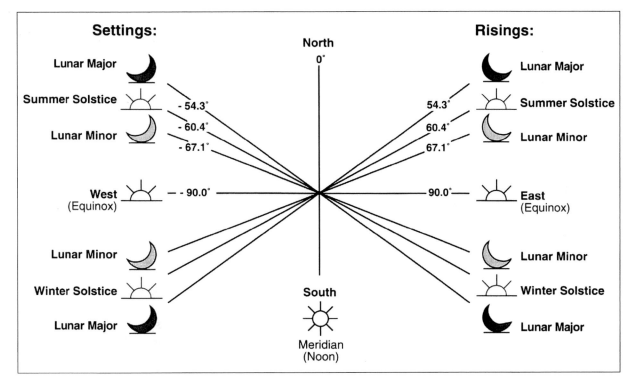

FIGURE 9.5. *Azimuths of the rising and setting of the sun and moon at the extremes and mid-positions of their cycles, at the latitude (36° north) of Chaco Canyon. The meridian passage of the sun is also indicated. The lunar extremes are the northern and southern limits of rising or setting at the major and minor standstills. (Fabian Schmid, Davis, Inc.; and Suzanne Samuels, By Design Graphics, Copyright © 1995 Solstice Project.)*

portion of the object to sight on and what time to sight it in its rising or setting, as well as identifying the exact time of a solstice or a standstill. Calculations for the latitude and environment of the Chaco region show the standard deviation developed from these sighting conditions and uncertainties: 0.5° in locating a solstice event; 0.5° in locating the minor standstill; and 0.7° in locating the major standstill (Sinclair and Sofaer 1993; see also Hawkins 1973: 287–288).

The Solstice Project surveyed the orientations of the fourteen largest buildings of the Chaco cultural region as ranked by room count (Powers, Gillespie, and Lekson 1983) (Figures 9.1 and 9.3, Table 9.1). The group comprises twelve rectangular and two crescent-shaped buildings that contained 115 to 695 rooms and were one to four stories high (Powers, Gillespie, and Lekson 1983). Ten buildings are located in the canyon, and four are located outside the canyon.

The buildings in the survey represent the Chacoans' most elaborate architecture. They include all of the large buildings in the canyon and the only outlying buildings that share the massive scale and impressive formality of the large buildings in the canyon (Lekson 1991; Roney 1992).[8]

All of the buildings in the Solstice Project's studies were developed between the late 800s and 1120s (Lekson 1984; Marshall *et al.* 1979; Powers, Gillespie, and Lekson 1983). Although the earlier buildings were modified and whole new buildings were constructed within this period, all the buildings that the Solstice Project surveyed were in use and most were being extensively worked on in the last and most intensive phase of Chacoan construction, from 1075 to about 1115 (Lekson 1984).

Six teams, working with the Solstice Project between 1984 and 1989, surveyed the orientations of most of the exterior walls of the twelve rectangular buildings. (The teams did not survey three

TABLE 9.1. *Sizes and Orientations of Major Chacoan Buildings. (Positive azimuths are east of north; negative azimuths are west of north.)*

Building	Number of Rooms	Area (m²)	Length of Principal Wall or Axis (m)	Orientations of:		
				Princ. Wall or Axis	Perp.	Diagonals
Pueblo Bonito	695	18,530	65	0.21° 0.14°	-89.79° ± 0.14°	
Chetro Ketl	580	23,395	140	69.60° ± 0.50°	-20.40° ± 0.50°	-86.4°
Aztec	405	15,030	120	62.47° ± 0.33°	-27.53° ± 0.33°	86.6° 37.2° -81.4° 24.8°
Pueblo del Arroyo	290	8,990	80	24.79° ± 0.25°	-65.21° ± 0.25°	-1.6° 49.9°
Kin Bineola	230	8,225	110	78.7° ± 3.2°	-11.3° ± 3.2°	-77.6° 54.2°
Peñasco Blanco	215	15,010	100	36.8° ± 1.3°	-53.2° ± 1.3°	
Wijiji	190	2,535	53	83.48° ± 0.15°	-6.52° ± 0.15°	-62.0° 49.2°
Salmon Ruin	175	8,320	130	65.75° ± 0.15°	-24.75° ± 0.15°	88.4° 43.3°
Una Vida	160	8,750	80	-35.18° ± 0.15°	54.82° ± 0.15°	
Hungo Pavi	150	8,025	90	-85.24° ±0.15°	4.76° ±0.15°	-61.4° 70.7°
Pueblo Pintado	135	5,935	70	69.90° ±0.15°	-20.10° ± 0.15°	31.4°
Kin Kletso	135	2,640	42	-65.82° ± 0.64°	24.18° ± 0.64°	87.38° -38.09°
Pueblo Alto	130	8,260	110	88.9° ±1.3°	-1.1° ± 1.3°	-68.6° 64.8°
Tsin Kletzin	115	3,552	40	89° ± 2°	-1° ± 2°	-66° 51°

© 1994 Solstice Project

NOTE: *Number of rooms and area from Powers et al. 1983: Table 41.*

short exterior walls of the rectangular buildings because the walls were too deteriorated.) The Solstice Project also surveyed the long back wall and the exterior corners of Peñasco Blanco, as well as the two halves of the exterior south wall and the primary interior wall of Pueblo Bonito, which approximately divides the plaza. In addition, the Solstice Project surveyed the dimensions of most of the exterior walls of the fourteen buildings. The teams established references at the sites by orienting to the sun, Venus, Sirius, or Polaris, or by tying to first- and second-order survey control stations.

Most of the walls are quite straight and in good condition at ground level and can be located within a few centimeters. Ten to thirty points were established along the walls and were measured in relation to the established references. These values were averaged to calculate the orientations of the walls. The Solstice Project was able to estimate based on multiple surveys of several walls that most of its measurements are accurate to within 0.25° of the orientation of the original walls. (Table 9.1 indicates where the survey was less accurate.)

The survey defined the orientations of the twelve rectangular buildings as either the direction of the longest wall (termed here the "principal" wall) or the perpendicular to this wall.[9] In all but one of the rectangular buildings, this perpendicular represents the "facing" direction of the building, the direction that crosses the large plaza. With respect to the crescent shaped buildings, the orientation of Pueblo Bonito is defined as the primary interior wall that approximately divides the plaza and the perpendicular to that wall, which corresponds closely in its orientation to that of a major exterior wall.[10] The orientation of Peñasco Blanco is defined by its symmetry as the line between the ends of the crescent and the perpendicular to this line (Figure 9.6).

The results of the survey show that the orientations of eleven of the fourteen major buildings are associated with one of the four solar or lunar azimuths on the sensible horizon (Tables 9.1 and 9.2, and Figure 9.6).[11] Three buildings (Pueblo Bonito, Pueblo Alto, and Tsin Kletzin) are associated with the cardinal directions (meridian and

equinox). One building (Aztec) is associated with the solstice azimuth. Five buildings (Chetro Ketl, Kin Kletso, Pueblo del Arroyo, Pueblo Pintado, and Salmon Ruin) are associated with the lunar minor standstill azimuth (Figure 9.7), and two buildings (Peñasco Blanco and Una Vida) are associated with the lunar major standstill.[12]

The orientations of the eleven major buildings that are associated with solar and lunar azimuths fall within 0.2° and 2.8° of the astronomical azimuths on the sensible horizon. Of these eleven, nine fall within 0.2° and 2.1° of the astronomical azimuths. The remaining two buildings, Chetro Ketl and Pueblo Pintado, are oriented respectively within 2.5° and 2.8° of the azimuth of the lunar minor standstill. (The wider differences in the orientations of these latter buildings from the lunar minor standstill are in the direction away from the solstice azimuth, which reinforces the conclusion that these buildings are associated with the moon rather than the sun.)

A number of factors (together or separately) could account for the divergence of the actual orientations of the major Chacoan buildings from the astronomical azimuths. These may include small errors in observation, surveying, and construction and a desire by the Chacoans to integrate into their astronomically oriented architecture symbolic relationships to significant topographic features and/or other major Chacoan buildings. (See for example the discussion in this essay of the solar-lunar regional pattern among the major Chacoan buildings.)[13]

The Solstice Project found that the eleven buildings that are oriented to astronomical events on the sensible horizon are also oriented to the same events on the visible horizon. The reason for this is that the topography introduces no significant variable in the observation of the rising or the setting astronomical events from these buildings. The divergence of the orientations of these buildings from the azimuths of astronomical events in one direction on the visible horizon (0.5° to 2.5°) is approximately the same as the divergence described above of their orientations from the azimuths of the same astronomical events on the

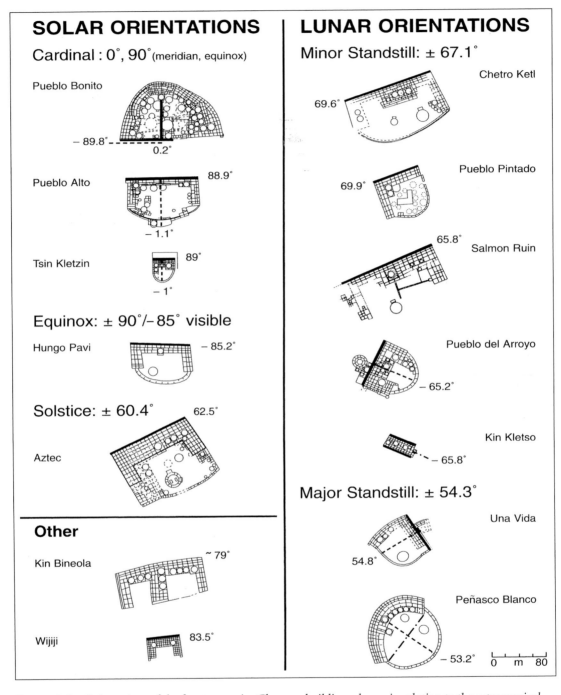

FIGURE 9.6. *Orientations of the fourteen major Chacoan buildings shown in relation to the astronomical azimuths on the sensible horizon. For one building, Hungo Pavi, the orientation to the equinox sunrise on the visible horizon also is indicated. (Suzanne Samuels, By Design Graphics, Copyright © 1995 Solstice Project.)*

	Principal Wall or Axis	Perpendicular	
Pueblo Bonito	0.2°	-89.8°	
Pueblo Alto	88.9°	-1.1°	} 0°, 90° Cardinal (meridian, equinox)
Tsin Kletzin	89.0°	-1.0°	
Hungo Pavi	-85.2°		} -85° Equinox/visible
Aztec	62.5°		} 60.4° Solstice
Peñasco Blanco		-53.2°	} 54.3° Lunar Major Standstill
Una Vida		54.8°	
Pueblo del Arroyo		-65.2°	
Kin Kletso	-65.8°		
Salmon Ruin	65.8°		} 67.1° Lunar Minor Standstill
Chetro Ketl	69.6°		
Pueblo Pintado	69.9°		
Wijiji	83.5°		
Kin Bineola	~ 79.0°		

© Solstice Project 1995

TABLE 9.2. *Orientations of Major Chacoan Buildings. (Positive azimuths are east of north; negative azimuths are west of north.)*

sensible horizon.[14] The differences between the orientations to the sensible and those to the visible horizon are so small as to not clearly indicate to which of these horizons the architects of Chaco oriented their buildings. The Solstice Project finds no evidence that the Chacoans were interested in making such a distinction in the case of eleven buildings.

Hungo Pavi, the twelfth building, appears to be oriented too far (4.8°) from the equinox rising or setting sun on the sensible horizon to qualify as an orientation associated with the solar azimuths on that horizon. It is, however, oriented to within one degree of the visible equinox sunrise.[15] Because of the topography, there is no corresponding visibility from Hungo Pavi to the equinox setting sun.

With respect to Wijiji and Kin Bineola, there appear to be no solar or lunar events associated with either the sensible or the visible horizon.[16] To conclude, orientation to the extremes and mid-positions of the solar and lunar cycles apparently played a significant role in the construction of the primary Chacoan architecture. No utilitarian reasons appear to explain the astronomic orientations of twelve of the fourteen major buildings.

Other researchers of prehistoric puebloan buildings report solar and lunar orientations and associations. At Hovenweep, in southern Utah, the orientations and locations of portholes of certain tower-like structures appear to be related to the solar cycle (Williamson 1984). Chimney Rock, an outlying Chacoan building in southern Colorado, appears to have been situated for its view of the major northern standstill moon rising between natural stone pillars, "chimney rocks" (Malville and Putnam 1989; Malville *et al.* 1991). The relationship of this building to the lunar major stand-still moon is underscored by the close correspondence of the tree-ring dates of its timbers with the occurrences of the lunar major standstill (1075 and 1094) at the peak of the Chacoan civilization. These findings in the outlying region of the Chacoan culture, as well as earlier find-ings of solar and lunar light markings in Chaco Canyon, support the phenomenon of solar and lunar orientations in the pri-mary Chacoan buildings.

SOLAR-LUNAR GEOMETRY INTERNAL TO THE MAJOR CHACOAN BUILDINGS

The Solstice Project's survey of the eleven rectangular major Chacoan buildings found strictly repeated internal diagonal angles and a correspondence between these angles and astronomy. The internal angles formed by the two diagonals and the long back walls of the rectangular buildings cluster in two groups (Figure 9.8A): sixteen angles in nine buildings are between 23° and 28°;[17] and six angles in four buildings are between 34° and 39°. (One of the buildings, Aztec, was con-structed first as a rectangular building with shorter side walls [Aztec I] that were extended in a later building stage [Aztec II] [Ahlstrom 1985]. It is of interest that when the side walls of Aztec I were extended to form Aztec II, the builders shifted from one pre-ferred angle to the other.)

At the latitude of Chaco, the angles between the lunar standstill azimuths on the sensible hori-zon and the east-west cardinal axis are 22.9° and 35.7°, respectively (Figure 9.8B). The correspon-dence between these angles of the solar-lunar relationships and the internal diagonal angles is intriguing. It suggests that the Chacoans may have favored these particular angles in order to incorpo-rate a geometry of the sun and moon in the internal organization of the buildings.[18]

FIGURE 9.7. *The moonrise seen through two doorways of Pueblo del Arroyo on April 10, 1990, when the moon rose at −67.5° on the visible horizon, close to the 67.1° azimuth of the lunar minor stand-still. Although we do not know whether an exterior wall, which is now deteriorated, blocked this view, the photograph illustrates the framing of the minor standstill moon by other exterior doorways and it conveys the perpendicular direction of the building toward the minor standstill moon (see Figure 9.6). (Photograph by Crawford MacCallum, Copyright © 1990 Solstice Project.)*

In addition, three rectangular buildings (Pue-blo Alto, Salmon Ruin, and Pueblo del Arroyo) are oriented on the sensible and visible horizons along one or both of their diagonals, as well as on their principal walls or perpendiculars, to the lunar minor standstill azimuth and to one of the cardi-nals (Table 9.1). The Chacoans may have intended the two phenomena—internal geometry and exter-nal orientation—to be so integrated that these three rectangular buildings would have both solar and lunar orientation.

A similar solar-lunar geometry appears to have guided the design of all of the major Chacoan

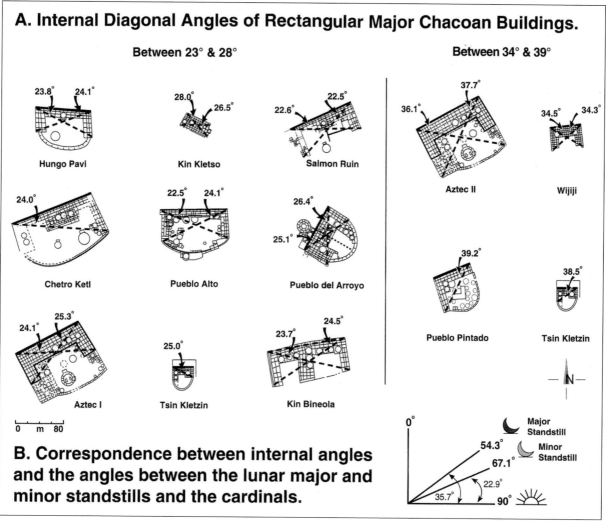

FIGURE 9.8. *(A) The eleven rectangular major Chacoan buildings, showing their diagonals and internal diagonal angles. (Two building phases at Aztec are shown.) (B) The correspondence of these angles to the angles between the lunar standstill azimuths and the cardinal directions. (Suzanne Samuels, By Design Graphics, Copyright © 1995 Solstice Project.)*

buildings (Sofaer 1994).[19] Furthermore, as with the three rectangular buildings discussed above, it appears that certain other of the major buildings also contain both solar and lunar orientations.

SOLAR-LUNAR REGIONAL PATTERN BETWEEN THE MAJOR CHACOAN BUILDINGS

Having seen that the Chacoans oriented and internally proportioned their major buildings in relationship to astronomy, the Solstice Project asked if the geographical relationships between the major buildings likewise expressed astronomical significance.

One scholar observed that four key central buildings are organized in a cardinal pattern (Fritz 1978). The line between Pueblo Alto and Tsin Kletzin is north-south; the line between Pueblo Bonito and Chetro Ketl is east-west. This work also showed that these cardinal interrelationships of four central buildings involved a symmetric patterning. The north-south line between Pueblo Alto and Tsin Kletzin evenly divides the east-west line between Pueblo Bonito and Chetro Ketl.

95

TABLE 9.3. *Astronomical Bearings between Astronomically Oriented Buildings. (Positive azimuths are east of north; negative azimuths are west of north.)*

Astronomically Oriented Buildings	Astronomical Bearings to Other Buildings			
	Buildings	Azimuth (degrees)	Differences between astronomical azimuth and interbuilding bearings (degrees)	Distance (km)
Cardinal Buildings associated azimuths 90°/0°				
Pueblo Bonito	Chetro Ketl	-88.7	-1.3	0.72
	Aztec	-2.2	2.2	86.3
Pueblo Alto	Tsin Kletzin	0.6	-0.6	3.7
	Aztec	-2.5	2.5	86.0
Hungo Pavi	- -	- -	- -	- -
Tsin Kletzin	Pueblo Alto	0.6	-0.6	3.7
	Aztec	-2.3	2.3	89.0
Solstice Building associated azimuth ±60.4°				
Aztec	- -	- -	- -	- -
Lunar Minor Buildings associated azimuth ±67.1°				
Chetro Ketl	Kin Bineola	69.3	-2.2	17.1
	Pueblo Pintado	-69.9	2.8	27.2
	Kin Kletso	-69.9	2.8	1.5
Pueblo del Arroyo	Hungo Pavi	-69.3	2.2	3.4
	Kin Bineola	67.8	-0.7	16.2
	Wijiji	-65.9	-1.2	8.4
Salmon Ruin	- -	- -	- -	- -
Pueblo Pintado	Chetro Ketl	-69.9	2.8	27.2
	Pueblo Bonito	-70.3	3.2	27.9
	Peñasco Blanco	-68.6	1.5	32.1
	Pueblo Alto	-68.0	0.9	27.8
	Kin Kletso	-69.9	2.8	28.7
Kin Kletso	Chetro Ketl	-69.9	2.8	1.5
	Pueblo Pintado	-69.9	2.8	28.7
	Wijiji	-64.5	-2.6	9.0
	Pueblo Alto	65.8	-1.3	1.3
	Kin Bineola	65.9	-1.2	16.0
	Hungo Pavi	-65.2	-1.9	4.0
Lunar Major Buildings associated azimuth ±54.3°				
Peñasco Blanco	Pueblo Bonito	-55.9	1.6	4.2
	Pueblo del Arroyo	-55.8	1.5	4.1
	Una Vida	-56.7	2.4	9.8
	Kin Bineola	55.0	0.7	14.3
Una Vida	Chetro Ketl	-51.3	-3.0	4.8
	Pueblo Bonito	-55.8	1.5	5.4
	Peñasco Blanco	-56.7	2.4	9.8
	Pueblo del Arroyo	-55.8	3.0	5.7
	Kin Kletso	-55.8	1.4	6.3

© Solstice Project 1995

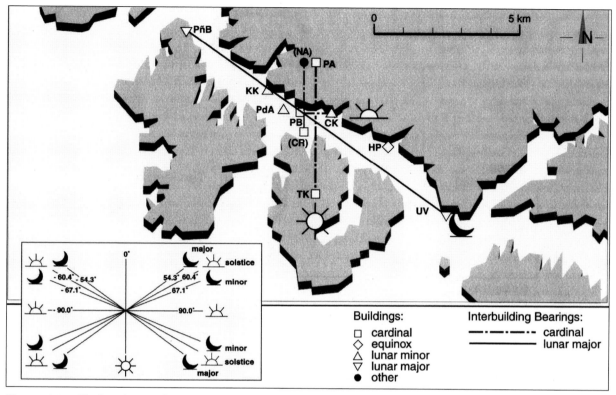

FIGURE 9.9. *The locations and orientations of the buildings in Chaco Canyon. The diagram shows the bearings between buildings that correlate with the orientations of individual buildings to the cardinal directions and to the lunar major standstill azimuths. (Fabian Schmid, Davis, Inc., Copyright © 1995 Solstice Project.)*

The Solstice Project found, in addition, that three of the four buildings involved in these cardinal interbuilding relationships are cardinal in their individual building orientations (Table 9.3; Figures 9.2, 9.9, and 9.10).[20] These findings suggest that the Chacoans coordinated the orientations and locations of several central buildings to form astronomical interbuilding relationships. The Project then asked if there were other such relationships between the major buildings.

As Table 9.3 and Figures 9.9 and 9.11 show, numerous bearings between thirteen of the fourteen major buildings align with the azimuths of the solar and lunar phenomena associated with the individual buildings.[21] Only one major building, Salmon Ruin, is not related in this manner to another building. In questioning the extent to which these astronomical interbuilding relationships were intentionally developed by the

Chacoans, the Solstice Project examined the pattern formed by them. In a manner similar to the central cardinal patterning, the bearings between the lunar-oriented buildings and other buildings appear to form lunar-based relationships that are symmetric about the north-south axis of the central complex (Figure 9.11).

The two isolated and remote outlying buildings, Pueblo Pintado and Kin Bineola, 27 km and 18 km, respectively, from the canyon center, are located on lines from the central complex that correspond to the bearings of the lunar minor standstill. As in the cardinal patterning, these lunar-based interbuilding relationships are underscored by the fact that they involve buildings that also are oriented individually to the lunar standstills (for one example see Figure 9.12a). Specifically, Chetro Ketl, Pueblo del Arroyo, and Kin Kletso—the three buildings in the central

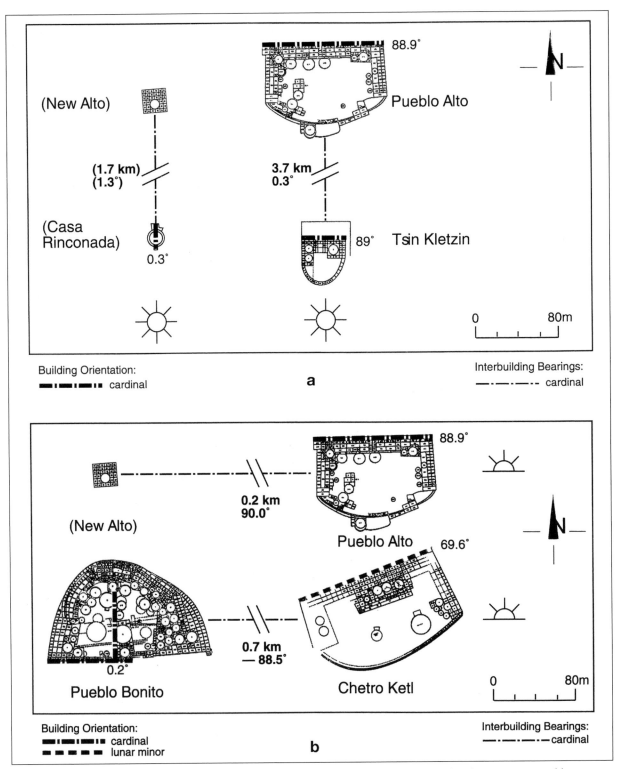

FIGURE 9.10. *The relationships between pairs of Chacoan buildings in the central complex that are connected by astronomical bearings: (a) north-south connections, and (b) east-west connections. (Fabian Schmid, Davis, Inc., Copyright © 1995 Solstice Project.)*

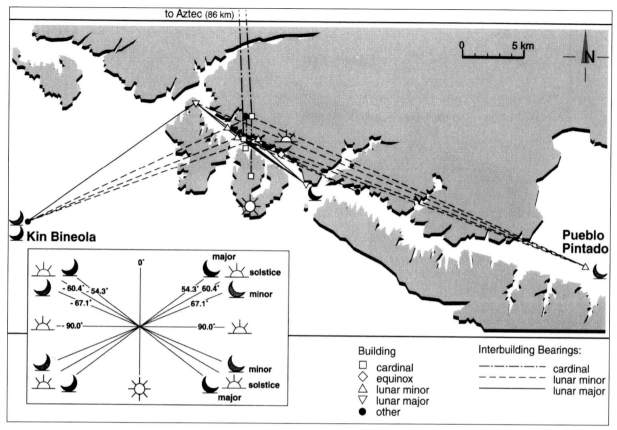

FIGURE 9.11. *The locations and orientations of twelve of the major Chacoan buildings, including Kin Bineola and Pueblo Pintado outside the canyon. The diagram shows bearings between buildings that correlate with the orientation of the individual buildings to the cardinal directions and the lunar major and minor standstill azimuths. (Fabian Schmid, Davis, Inc., Copyright © 1995 Solstice Project.)*

complex that are oriented to the lunar minor standstill—also are related to Pueblo Pintado and Kin Bineola on bearings oriented to the lunar minor standstill. It is of interest that Pueblo Pintado also is oriented to the lunar minor standstill (Figures 9.11 and 9.12a).[22] In addition, two major buildings, Wijiji and Hungo Pavi, located outside of the central complex but within the canyon, also are on the bearing from the central complex to Pueblo Pintado and to the lunar minor standstill (Figure 9.11).[23]

The relationship of the central complex to Pueblo Pintado (southeast of the canyon) is to the rising of the southern minor standstill moon; the relationship of the central canyon complex to Kin Bineola (southwest of the canyon) is to the setting of this same moon. Thus the north-south axis of

the central complex is the axis of symmetry of this moon's rising, meridian passage, and setting, as well as the axis of the ceremonial center and of the relationships of these significant outlying structures to that center.

It is of further note that Pueblo Pintado and Kin Bineola are regarded as having particularly significant relationships with the buildings in the canyon. One archaeologist reports that these two buildings are more like the canyon buildings than they are like other outlying buildings, and he suggests, because of their positions to the southeast and southwest of the canyon, that they could be viewed as the "gateway communities" (Michael P. Marshall, personal communication 1990).

This lunar-based symmetrical patterning about the north-south axis of the central ceremonial

99

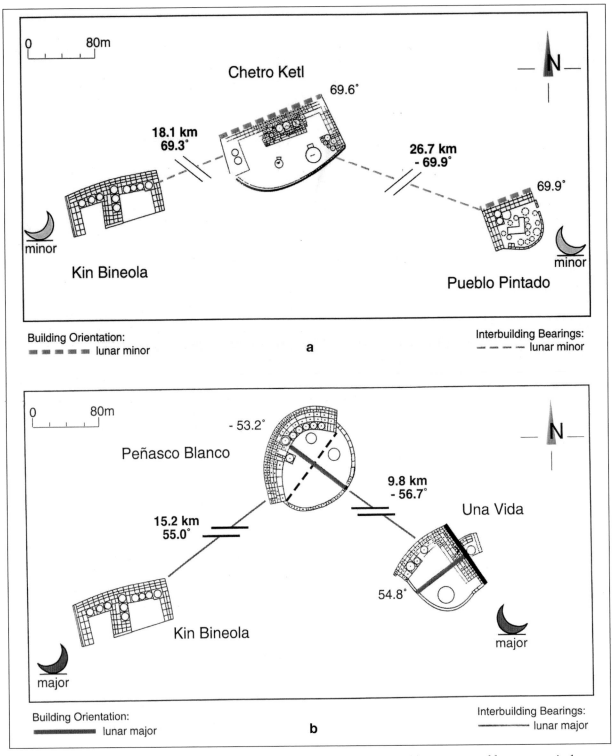

FIGURE 9.12. *The relationships between two groups of three major Chacoan buildings connected by astronomical bearings aligned to (a) the lunar minor standstill and (b) the lunar major standstill. (Fabian Schmid, Davis, Inc., Copyright © 1995 Solstice Project.)*

complex also is expressed in the relationships of the lunar major–oriented buildings, Una Vida and Peñasco Blanco, to that complex (Figure 9.9). Without knowing the astronomical associations of these buildings, other scholars had observed the symmetrical relationship of Una Vida and Peñasco Blanco to the north-south axis, as described above, between two major buildings in the central complex, Pueblo Alto and Tsin Kletzin; and one of these scholars described this relationship as, along with the cardinal relationships of the central complex, "establishing the fundamental symmetry of the core development of Chaco Canyon" (Fritz 1978; Stein and Lekson 1992).

From the central complex, bearings to the major standstill moon are also the bearings to Una Vida and Peñasco Blanco, the only major Chacoan buildings that are oriented to the lunar major standstill (Figure 9.12b). This correspondence of the interbuilding relationships with the individual building orientations is again what is found with the cardinal and lunar minor relationships of the major buildings. Here it also is striking that the two buildings are equidistant from the north-south axis of the central complex. It is of further interest that the bearing from Peñasco Blanco to Kin Bineola also corresponds with the bearing to the lunar major standstill (Figure 9.11).[24] Una Vida, Peñasco Blanco, and Kin Bineola, along with Pueblo Bonito, share the earliest dates among the major Chacoan buildings (Lekson 1984; Marshall *et al.* 1979).

Thus, from the central complex of Chaco Canyon, in the year of the major standstill moon, there was a relationship to that moon, as it rose farthest south in its full cycle, that also incorporated a relationship to Una Vida; and, in that same year, as the moon made its excursion to setting farthest north in its full cycle, it was on a bearing from the central complex that incorporated a relationship with Peñasco Blanco. Furthermore, in that year, the southern major standstill moon that rose on the bearing from Peñasco Blanco to Una Vida and the central complex would set on the bearing

from Peñasco Blanco to the outlying major building, Kin Bineola. This phenomenon may have been intended to draw Kin Bineola into a lunar major relationship with Peñasco Blanco and with Peñasco Blanco's lunar major connection with Una Vida and the central canyon complex.[25]

At the other end of the lunar standstill cycle, nine to ten years later, in the year of the minor standstill moon, two outlying buildings, Kin Bineola and Pueblo Pintado, would be drawn into relationship with the central complex by their locations on bearings from the central complex that are to the rising and setting of the southern minor standstill moon.

Finally, in the face of the evidence that the Chacoans oriented and proportioned their major buildings in relationship to the solar and lunar cycles—and also interrelated their cardinally oriented buildings in a cardinal and symmetrical pattern—it is difficult to dismiss as coincidental the lunar-based interbuilding relationships, which are based on the same principles.[26] The recurring correlation of the interbuilding lines with the astronomical phenomena associated with the individual Chacoan buildings, and the centrally and symmetrically organized design of these lines, suggest that the Chacoan culture coordinated the locations and orientations of many of its major buildings to form an interbuilding regional pattern that commemorates and integrates the cycles of the sun and the moon.[27]

Most of the buildings related by astronomical interbuilding lines are not intervisible. In general, this is because the canyon and other topographic features block the views between the buildings, especially those related over long distances. Thus the astronomical interbuilding lines could not have been, in general, used for astronomical observations or predictions.[28] It is of interest that the Chacoan roads, which are typified by their rigorous straight course, frequently appear to ignore topographic obstacles and connect sites that are great distances apart and are not intervisible.

CONSIDERATION OF THE EVOLUTION OF ASTRONOMICAL EXPRESSION IN CHACOAN ARCHITECTURE

The evidence of a conscious effort by the Chacoans to orient and interrelate their buildings on astronomical bearings raises a number of questions for further study. Were the building locations selected because they fell on astronomical bearings from other buildings? Were the interbuilding bearings developed from a plan? Were some buildings originally located for reasons other than astronomy and later drawn into the astronomical regional pattern?

It will probably never be known how great a role astronomy played in the decisions regarding the placement of the Chacoan buildings. Nor does it appear possible to know the extent of planning that preceded the development of the astronomical expressions in Chacoan architecture.[29] The data presently available on the chronology of the construction of the major buildings, however, do provide some insight into the history of the development of astronomical orientations and interrelationships among the major Chacoan buildings.

These data show that astronomical orientation appears to have played a part in Chacoan architecture from the earliest to the latest phases of its construction. Pueblo Bonito's north-south axis was incorporated in a major interior wall in the building's earliest design in the late 800s (Stein, Suiter, and Ford 1997), and this north-south axis was extended and elaborated in the construction of the primary interior wall during the building's last phase of construction, in the late 1090s. The cardinal orientations of Pueblo Alto and Tsin Kletzin were developed in the early 1000s and the early 1100s, respectively. The lunar orientations were developed from the mid 900s (in Una Vida) through the early 1100s (in Kin Kletso).

Available data on the evolution of individual buildings show that, for most of the fourteen major buildings, the walls that are revealed today—and that were the subject of this study—are the original walls of these buildings, or that they

follow closely the orientation of the buildings' prior walls. These data indicate that the orientation of two of the fourteen buildings changed significantly from one building phase to another.

It is also of interest that three of the four buildings in which the earliest dates were found among the fourteen major buildings (Peñasco Blanco, Una Vida, and Kin Bineola) are involved in the lunar major standstill interbuilding bearings. (Curiously, Peñasco Blanco is one of the two buildings that shifted from an earlier orientation [of −67°, in 900–915] to a later orientation [of −53.2° in 1050–1065].)

With further dating information it may be possible to know more of the evolution of astronomical expression in the Chacoan buildings. Such information could also shed light on the intriguing possibility that there may be, as there was found to be at Chimney Rock, correlations between the building phases of the major Chacoan buildings and the astronomical cycles (Malville and Putnam 1989).

SPECULATIONS ON THE CHACOANS' EXPERIENCE

Many of the major buildings appear to incorporate interesting views and experiences of the sun and moon at the extremes and mid-positions of their cycles. For example, each day at meridian passage of the sun, the mid-wall which approximately divides the massive structure of Pueblo Bonito casts no shadow. Similarly, the middle of the sun's yearly passage is marked at Pueblo Bonito as the equinox sun is seen rising and setting closely in line with the western half of its south wall. Thus, the middle of the sun's daily and yearly journeys are visibly in alignment with the major features of this building, which is at the middle of the Chacoan world.[30]

From many of the other major buildings, the sun and moon at the extreme positions of their cycles would be seen rising and/or setting along the long back walls or across the plazas at angles perpendicular to the back walls. In buildings oriented in their facing directions to the lunar standstill azimuths, the rising or setting moon, near its extremes, would be framed strikingly by the doorways (Figure 9.7).

Also visually compelling would have been the view from Peñasco Blanco of the moon rising at the major standstill position. This building is located 5.4 km northwest of Pueblo Bonito near the top of West Mesa. From it, one would view the southern major standstill moon rising in line with the mid-axis of the building's crescent, and also on a bearing to Pueblo Bonito and to the central complex of the canyon. The bearing would appear to continue through the valley of the canyon to the rising moon on the horizon. This event marked the time when the moon rises farthest south in its full cycle, once every eighteen to nineteen years.

This dramatic view of the major standstill moonrise also embodied astronomical and symmetrical relationships to nonvisible objects. Out of sight, but on the alignment between the viewer at Peñasco Blanco and the rising moon, is Una Vida, the one other of the Chacoan buildings that is oriented to the major standstill moon. Some viewers would likely have known of this nonvisible building's position on the bearing from Peñasco Blanco to the major rising moon, and they may also have known of Una Vida's and Peñasco Blanco's symmetrical relationship with the north-south axis of the central complex—in other words, that the two buildings are located the same distance from the north-south axis. Seeing the southern major standstill moon set over the mesa rise behind Peñasco Blanco would have conveyed to some Chacoans that as it set, out of view, on the sensible horizon it was on a bearing with Kin Bineola, out of view, 14.3 kilometers to the southwest. Thus the experience of viewing the moon rising and setting at its southern major standstill from Peñasco Blanco would have involved seeing certain visible—and knowing certain nonvisible—aspects of the building's relationships with astronomy and with other major Chacoan buildings.

In the sculptured topography of the southern Rockies, at a location 150 kilometers north of Chaco, the Chacoans witnessed a spectacular view of the moonrise at its major standstill. From their building situated high on an outcrop at Chimney Rock, once every eighteen to nineteen years, the Chacoans watched the moon rise between two nearby massive stone pillars.

Thus, while certain aspects of Chacoan architecture embed relationships on astronomical bearings to nonvisible objects, others appear to have been designed and/or located to frame, or to align to, bold displays of astronomy. Furthermore, some Chacoan astronomical expressions are on bearings that ignore topographic features, while others use topography dramatically to reinforce the visual effects of the architectural alignments to the sun and moon.

CONCLUDING DISCUSSION

Peoples throughout history and throughout the world have sought the synchronization and integration of the solar and lunar cycles. For example, in times and places not so remote from Chaco, the Mayas of Mesoamerica recognized the 19-year metonic cycle—the relationship of the phase cycles of the moon to the solar cycle—and noted elaborately, in the *Dresden Codex,* the pattern of lunar eclipses (Aveni 1980).[31] The Hopi, a Pueblo people living today in Arizona, are known to have synchronized the cycles of the sun and moon over a two-to-three-year period in the scheduling of their ceremonial cycle (McCluskey 1977). At Zuñi Pueblo in northwestern New Mexico, the joining of Father Sun and Mother Moon is sought constantly in the timing of ceremonies (Tedlock 1983).

Each of the Chacoan expressions of solar and lunar cosmology contains within it this integration of the sun and the moon. For example, at the three-slab site on Fajada Butte, the sunlight in a daggerlike form penetrates the center of the large spiral at summer solstice near midday, the highest part of its cycle (Sofaer, Zinser, and Sinclair 1979); and, as though in complement to this, the moon's shadow crosses the spiral center at the lowest point of its cycle, the minor standstill (Sofaer, Sinclair, and Doggett 1982). In the same way, the outer edges of the spiral are marked by the sun in light patterns at winter solstice, and the moon's shadow at its maximum extreme is tangent to the left edge.

This integration of the sun and the moon is in the three expressions of solar-lunar cosmology in Chacoan architecture. Five major buildings commemorate the solar cycle: three in their cardinal orientations, one in its equinox orientation, and one in its solstice orientation. Seven of the other nine major buildings commemorate the lunar standstills: five the minor standstill, and two the major standstill. And the overall patterning of the buildings joins the two sets of lunar-oriented buildings into relationship with the cardinal-solar center in a symmetrically organized design. The geometry of the rectangular buildings again expresses the joining of sun and moon; the internal angles related to the cardinal and lunar azimuths bring a consciousness of each of these cycles into the layout of the buildings.

Commemoration of these recurring cycles appears to have been a primary purpose of the Chaco phenomenon. Many people must have been involved over generations in the planning, development, and maintenance of the massive Chacoan constructions. The work may have been accomplished in relatively short periods of time (Lekson 1984) and perhaps in episodes timed to the sun and moon. This activity would have unified the Chacoan society with the recurring rhythms of the sun and moon in their movements about that central ceremonial place, Chaco Canyon.[32]

There are many parallels to the cosmological patterning of the Chacoan culture in the architectural developments of the Mesoamerican cultures. These developments occurred in the region to the south of Chaco, for several centuries before, during, and after Chaco's florescence.

It is observed that "the coordination of space and time in the Mesoamerican cosmology found its expression in the orientations of pyramids and architectural complexes" (Broda 1982) and in the relationships of these complexes to outlying topography and buildings (Broda 1993). Ceremonies related to the dead and timed to the astronomical cycles occurred in Mesoamerican centers (Broda 1982). In central structures of the ceremonial complexes, light markings commemorated the zenith passage of the sun (Aveni 1980). Certain of the ceremonial centers were organized on axes close to the cardinal directions (Aveni 1980; Broda 1982). It is stated that cosmological expression in Mesoamerica "reached an astonishing degree of elaboration and perfection," and that its role was "to create an enduring system of order encompassing human society as well as the universe" (Broda 1993). A Mesoamerican archaeoastronomer comments that "a principle of cosmic harmony pervaded all of existence in Mesoamerican thought" (Aveni 1980).

The parallels between Mesoamerica and Chaco illustrate that the Chacoan and Mesoamerican peoples shared common cultural concerns. In addition, the several objects of Mesoamerican origin that were found in Chacoan buildings indicate that the Chacoans had some contact with Mesoamerica through trade.

In the complex cosmologies of the historic Pueblo peoples, descendants of the Chaocans, there is a rich interplay of the sun and moon.[33] Time and space are integrated in the marking of directions that order the ceremonial structures and dances, and in the timing of ceremonies to the cycle of the sun and the phases of the moon. The sun and the moon are related to birth, life, and death.[34] Commemoration of their cycles occurs on some ceremonial occasions in shadow-and-light patterns. For instance, sunlight or moonlight striking ceremonial objects or walls of ceremonial buildings may mark the solstices, as well as the meridian passage of the solstice sun and the full moon, and time the beginning and ending of rituals.

In many Pueblo traditions, the people emerged in the north from the worlds below and traveled to the south in search of the sacred middle place. The joining of the cardinal and solstice directions with the nadir and the zenith frequently defines, in Pueblo ceremony and myth, that sacred middle place. It is a center around which the recurring solar and lunar cycles revolve. Chaco Canyon may have been such a center place and a place of mediation and transition between these cycles and between the worlds of the living and the dead (F. Eggan, personal communication 1990).[35]

For the Chacoans, some ceremonies commemorating the sun and the moon must have been conducted in relatively private settings, while others would have been conducted in public and monumental settings. A site such as the three-slab site would have been visited probably by no more than two or three individuals, who were no doubt highly initiated, specialized, and prepared for witnessing the light markings.[36] By contrast, the buildings would have been visited by thousands of people participating in ceremony.

The solar and lunar cosmology encoded in the Chacoans' massive architecture — through the buildings' orientations, internal geometry, and geographic relationships — unified the Chacoan people with each other and with the cosmos. This order is complex and stretches across vast reaches of the sky, the desert, and time. It is to be held in the mind's eye, the one that sees into and beyond natural phenomena to a sacred order. The Chacoans transformed an arid empty space into a reach of the mind.

ACKNOWLEDGMENTS

The generous help and disciplined work of many individuals made possible the research presented here. We are particularly indebted to Rolf M. Sinclair (National Science Foundation) for his help in the rigorous collection and reduction of large amounts of data, for his thoughtful analysis of naked-eye astronomical observations, and for his help in organizing numerous and complicated surveying trips. We most especially appreciate his dedication to pursuing the truth through thousands of hours of discussion and analysis of the Chacoan material.

We wish to express a special thanks to John Stein (Navajo Nation) for sharing with us, over fifteen years, his many insights to the nonmaterial side of the Chaco phenomenon. To our knowledge, John was the first scholar to speak of Chacoan architecture as "a metaphoric language" and to express the view that the Chacoan roads were built for purposes other than utilitarian functions. He early on saw geometric complexity in the individual buildings and their interrelationships.

Phillip Johnson Tuwaletstiwa (Hopi Tribe), the late Fred Eggan (Santa Fe), and Jay Miller (Chicago) helped us move from the confines of an engineering and surveying perspective to think of the rich and complex dualities of Pueblo cosmology and their analogies with the Chacoan developments. The late Alfonso Ortiz (University of New Mexico), when first viewing our material on the patterning of Chacoan roads and major buildings, observed that these complex constructions were developed as though they were to be seen from above.

Tuwaletstiwa was also responsible for initiating the surveys that form the basis of our studies. Most of the surveys were carried out by Richard Cohen (National Geodetic Survey), assisted by Tuwaletstiwa, and supplemented by William Stone (National Geodetic Survey). Other surveys were conducted by Robert and Helen Hughes and E. C. Saxton (Sandia National Laboratories); C. Donald Encinias and William Kuntz, with the cooperation of Basil Pouls (Koogle and Pouls Engineering, Inc.); James Crowl and associates (Rio Grande Surveying Service); Scott Andrae (La Plata, NM), assisted by Dabney Ford (National Park Service) and Reid Williams (Rensselaer Polytechnic Institute); and William Mahnke (Farmington, NM).

Michael Marshall (Cibola Research Consultants) shared with us his comprehensive knowledge of Chacoan archaeology, and his keen insights into the Chacoans' use of topography. Stephen Lekson's (University of Colorado) extensive knowledge of the major buildings in Chaco Canyon was invaluable to us. We also benefited from many helpful discussions regarding Chacoan archaeology with John Roney (Bureau of Land Management), Dabney Ford, and Thomas Windes (National Park Service). LeRoy Doggett (U.S. Naval Observatory), Bradley Schaefer (Goddard Space Flight Center), and the late Gerald Hawkins (Washington, D.C.) assisted with a number of points concerning the naked-eye astronomical observations. Crawford MacCallum (University of New Mexico) photographed the moonrise under unusual lighting conditions.

We thank J. McKim Malville (University of Colorado) for a number of interesting discussions and a tour of several northern prehistoric puebloan building complexes, where he has identified astronomical building orientations and interbuilding relationships, and Murray Gell-Mann (Santa Fe Institute) for joining us for a tour of several major Chacoan buildings and contributing stimulating and encouraging comments on our early analysis of possible astronomical significance in the Chacoan architecture.

William Byler (Washington, D.C.) once more gave us extremely generous and thoughtful assistance in editing the final manuscript that was published in Morrow and Price's 1997 edited volume by the University of New Mexico Press. The graphics were prepared by Suzanne Samuels (By Design Graphics) and by Fabian Schmid with the cooperation of Davis, Inc. (Washington, D.C.).

The fieldwork in Chaco Canyon was done with the helpful cooperation of former Superintendents Thomas Vaughan and Larry Belli, with the thoughtful guidance of National Park Service archaeologist Dabney Ford, and with the generous assistance of many other staff members of Chaco Culture National Historical Park.

NOTES

1. For convenience the reader can find most of these papers on the Solstice Project's website: *www.solsticeproject.org.*

2. The long back walls of five of the ten major buildings located in Chaco Canyon are somewhat parallel to local segments of the north canyon wall. Since there are innumerable locations along this canyon wall where significantly different orientations occur and where these buildings could have been placed, this approximate parallel relationship does not appear to have been a constraint on the orientations of these five buildings.

3. In the literature of Chacoan studies, we find one suggestion of a utilitarian reason for the location of the major buildings, and it applies to only one building. Specifically, it has been suggested that Tsin Kletzin was placed to optimize the direct sight lines to six other buildings (Lekson 1984: 231). A suggestion by Judge (1989) that three major buildings "functioned primarily as storage sites to accompany resource pooling and redistribution within the drainage systems they 'controlled'" locates them only generally.

4. For an example of a nonutilitarian Chacoan road, see Dabney Ford's finding of a road connecting the canyon floor with the three-slab site on Fajada Butte (Ford 1993).

5. In addition, the relationship of Pueblo Bonito's design to the solar cycle appears to be symbolically represented in a petroglyph on Fajada Butte in Chaco Canyon (Sofaer and Sinclair 1989: 499; Sofaer 2006).

6. "Sensible horizon" describes the circle bounding that part of the earth's surface if no irregularities or obstructions are present. "Visible horizon" describes the horizon that is actually seen, taking obstructions, if any, into account.

7. It would seem unlikely that the Chacoans, who incorporated cardinal orientations in their architecture, and who also marked the equinoxes and meridian passage in light markings, did not associate the north-south axis with the sun's meridian passage and the east-west axis with the sun's rising and setting positions at equinox.

8. See also Lekson 1991: "Using intrinsic criteria, one could argue that only the Big Four (Salmon Ruin, Aztec, Pueblo Pintado, and Kin Bineola)...were identical to Pueblo Bonito and Chetro Ketl." Eventually, the Solstice Project will also study the "medium-size" (Powers *et al.* 1983) and the more remote Chacoan buildings for possible astronomical significance.

9. In most cases the longest wall is obvious. For the orientation of Pueblo Pintado, values were taken for the longer of the two walls and the perpendicular to it. For Kin Kletso, the orientations of the two long walls of equal length, which differed in orientation by only 0.8°, were averaged. Kin Bineola's principal wall is not a straight wall, but three sections, which vary by several degrees. The sections were averaged in the value given here, and the error quoted (±3°) reflects the differences in the sections.

10. The Solstice Project notes that other scholars have described the cardinal orientation of Pueblo Bonito by the direction of this primary interior wall and the direction of the western half of the south wall (Williamson *et al.* 1975, 1977). The eastern half of the south wall, which is not perpendicular to the primary interior wall and is oriented to 85.4°, is a curious departure from these perpendicular relationships.

11. The orientation of Hungo Pavi as reported here corrects an error in an earlier paper (Sofaer, Sinclair, and Donahue 1991). The orientations of nine other major Chacoan buildings are also reported here with slightly different values than those reported in the earlier paper. These changes are the result of certain refinements in a further reduction of the Solstice Project's survey data. The changes, unlike in the case of Hungo Pavi, are so slight (from 0.1° to 0.7°) that they do not affect the conclusions.

12. It is of interest that a unique and extensive construction of the Chacoan culture, the Chetro Ketl "field," which is a grid of low walls covering more than twice the land area of the largest Chacoan building, appears also to be oriented to the azimuth of the lunar minor standstill. This construction was reported to have an orientation of −67° (Loose and Lyons 1977). It should be further noted that the Solstice Project's survey found that the orientation of the perpendicular of Kin Klizhin, a tower kiva located 10 km from Chaco Canyon, is −65°, an azimuth also close to the azimuth of the lunar minor standstill.

13. In certain of the Solstice Project's earlier studies of Chacoan constructions, an emphasis was given to substantiating claims of accurate alignments. The author believes that this focus sometimes blinded us in our search for the significance of the orientations and relationships developed by this prehistoric and traditional society, to whom symbolic incorporation of astronomical relationships would have been at least as important as the expression of optimal accuracy. In addition, in several instances, the Project's studies have shown that alignments (such as the north orientation of the Great North Road) are adjusted off of precise astronomical direction in order to incorporate other symbolic relationships (Sofaer, Marshall, and Sinclair 1989).

14. The preliminary results of the Solstice Project's study of elevated horizons that are near certain of the major buildings show that from eight of these eleven buildings both the rising and setting astronomical events occur within 1° to 3° of the building orientations.

15. The preliminary results of the Solstice Project's current study show that none of the other thirteen buildings is oriented, as Hungo Pavi is, to an astronomical event on only the visible and not the sensible horizon.

16. The Solstice Project finds that the orientation of Wijiji, which is approximately 6.5° off of the cardinal directions, is also close to the orientation of New Alto, Aztec East, and the east and north walls of the great kiva of Pueblo Bonito, as well as the orientation of several interbuilding relationships. Although there is no obvious astronomical reason for the selection of this azimuth for building orientations and interrelationships, its repetition indicates that it may have been significant to the Chacoans. In addition, at Wijiji at winter solstice the sun is seen rising in a crevice on the horizon (Malville 2005:75). The Solstice Project survey shows that the alignment from Wijiji to this event is also the diagonal of the building. Other instances of astronomical orientation of the diagonals of the buildings are discussed in the next section of this chapter.

17. Because of the deterioration of one of its short walls, Chetro Ketl has only one measurable diagonal angle.

18. The Chacoans may have had additional reasons to consistently choose angles of approximately 23° and 36°. It has been suggested that these angles were also used by a Mesoamerican culture (Clancy 1994; Harrison 1994).

   It is of interest that only at locations close to the latitude of Chaco Canyon (*i.e.,* 36°) do the angles of 23° and 36° correspond with the relationships of the cardinal directions and the lunar major and minor standstill azimuths. In addition, at the latitude of 36° at solar noon on equinox day, the shadow of a stick or other vertical object cast on a flat surface forms a right angle triangle that has the internal angles of 36° and 54°. The correspondence of the internal angles of the major Chacoan buildings with angles apparently favored by a Mesoamerican culture, as well as with the angles evident in the solar and lunar astronomy that occurs only close to the latitude of Chaco, raises intriguing questions. It may be that Chaco Canyon was selected as the place, within the broader cultural region of Mesoamerica, where the relationships of the sun and the earth, and the sun and the moon, could be expressed in geometric relationships that were considered particularly significant.

Of further interest is one archaeologist's discussion of the location of Chaco Canyon and Casas Grandes, a postclassic Mesoamerican site, on the same meridian. He suggests that this correspondence may have been an intentional aspect of the locating of Casas Grandes (Lekson 1996). Casas Grandes is 630 km south of Chaco Canyon.

19. The Solstice Project's further study of the internal design of the major Chacoan buildings suggests that one of the solar-lunar angles found in the rectangular buildings, 36°, is also incorporated in the design of three other major buildings (Pueblo Bonito, Peñasco Blanco, and Una Vida) and that Kin Bineola's design (like Aztec I and II) incorporates 36° as well as 24°. In addition, in several of these buildings the solar-lunar geometry is combined with orientational relationships to both the sun and the moon (Sofaer 1994). It also is of interest that three great kivas in Chaco Canyon are organized in geometric patterns of near-perfect squares and circles. This further geometric study of Chacoan architecture will be presented in work that is in preparation by the Solstice Project.

20. The Solstice Project also found that cardinal interbuilding lines relate two minor buildings located in the central canyon to each other and to one of the major central buildings involved in the central cardinal patterning. The line between Casa Rinconada, the cardinally oriented great kiva, and New Alto aligns closely with the north-south axis of Casa Rinconada, and New Alto lies directly west of the cardinally oriented Pueblo Alto (Figures 9.2, 9.3, 9.9, and 9.10). An internal feature of Casa Rinconada appears to mark the kiva's north-south relationship with New Alto. The south stairway of Casa Rinconada is positioned slightly off the axis of symmetry of the kiva, and this stairway is also offset in the south doorway. The effect of the offset placement of this stairway is that from its center one sees New Alto over the center of the north doorway on a bearing of 1.3°. (Although the construction of Casa Rinconada was completed before the construction of New Alto, it is possible that the position of the stairway within the south doorway of Casa Rinconada was modified at the time of New Alto's construction.)

Three long, low walls extending from Pueblo Alto (surveyed by the Solstice Project) are also cardinally oriented, and they appear to further elaborate the cardinal pattern of the central complex (Windes 1987).

21. The astronomical interbuilding bearings shown in Table 9.3 and in Figures 9.9 and 9.11 are defined as the bearings between two buildings that align (within 3°) with the rising or setting azimuths of the astronomical phenomena associated with one of the two buildings.

The Solstice Project identified the locations of the fourteen major buildings from the coordinates of the 7.5' topographic survey maps of the U.S. Geological Survey. The relative locations of certain of the central buildings were confirmed by direct surveying and by the use of existing aerial photography. The bearings of the interbuilding lines were taken from the estimated centers of the buildings. (The close relationship of two very large buildings, Pueblo Bonito and Chetro Ketl, introduced the only uncertainty. In this case, however, it was observed that each point in Chetro Ketl is due east of each point in Pueblo Bonito.) The relative locations of the buildings could be identified to within 15 m on the maps. The Solstice Project estimates that its measurements have a typical uncertainty in the bearing of an interbuilding line of 0° 12' at an average separation of 4.7 km for the ten buildings within the canyon, and much less uncertainty in the bearings of interbuilding lines extending outside the canyon.

22. The orientation of the perpendicular of Pueblo Pintado is to the azimuth that corresponds with a bearing to Salmon Ruin, 85 km from Pueblo Pintado; furthermore, the azimuth of the orientation of the perpendicular of Salmon Ruin also corresponds with this bearing. Perhaps these relationships were deliberately developed by the Chacoans to join two outlying major buildings that are oriented to the minor standstill moon on a bearing perpendicular to the azimuth of the minor standstill moon and to draw Salmon Ruin into connection with the central complex of Chaco Canyon, to which Pueblo Pintado is related by lunar minor standstill relationships (as is suggested elsewhere in this chapter).

23. It is of interest that the two other Chacoan constructions, the Chetro Ketl "field" and Kin Klizhin (a tower kiva), that are oriented to the lunar minor standstill are also on the lunar minor standstill interbuilding bearings from the central complex to Kin Bineola and to Pueblo Pintado, respectively (see note 12).

24. The Solstice Project's preliminary investigations of several C-shaped, low-walled structures (Windes

1978) and three sets of cairns located in and near Chaco Canyon show that the bearings between these sites are oriented to the lunar major standstill. It is also of interest that several recent findings by others suggest astronomical relationships among sites within prehistoric pueblo building complexes, including one Chacoan building complex, in southwestern Colorado (Malville *et al.* 1991; Malville and Putnam 1989).

25. It is of interest that Ron Sutcliffe documented another interbuilding bearing: Peñasco Blanco to Casa Rinconada on the alignment to the rising of the southern major standstill moon, June 11, 2006.

26. Certain of the astronomical interbuilding relationships within the canyon, such as that between Una Vida and Peñasco Blanco, appear to correspond roughly with the topography of the canyon. While this correspondence suggests the possibility that the relationship between these buildings could have fallen into lunar alignment by coincidence, it does not explain the other interlocking aspects of these buildings, which suggest an intentional marking of the lunar major relationship between them. The relationships of the central complex to Pueblo Pintado and Kin Bineola on the lunar minor bearings are not affected by the canyon topography because these buildings are located beyond the canyon. The lunar minor relationships of the central complex to Hungo Pavi and Wijiji could have been affected in part by accommodation to the canyon topography. This would not discount the possibility that these relationships had lunar significance for the Chacoans.

27. Although the Solstice Project cannot be certain that all of the astronomical interbuilding bearings that are shown in Table 9.3 and Figure 9.11 were intentionally developed by the Chacoans, it seems important at this stage in our study to present all the interbuilding bearings that meet the criterion described above (see note 21).

One astronomical interbuilding bearing which has not been discussed in the text deserves particular note. Aztec, 86 km north of Pueblo Bonito, is located on a bearing from the central complex of Chaco that could have been regarded by the Chacoans as a continuation of the north-south axis of the central buildings and their interbuilding relationships (Table 9.3 and Figure 9.11). Certain analysis suggests that the north-south bearing between Chaco and Aztec had particular significance to the Chacoans. Aztec, itself a massive architectural complex, is regarded as an important late center of the Chacoan culture. An architectural study shows that Aztec appears to be "modeled on standards fixed in Pueblo Bonito" (Stein and McKenna 1988). An author of this latter study further notes that the core activity of the Chacoan culture moved in the late 1100s from Chaco Canyon to Aztec (Fowler and Stein 1992), and that this center maintained an active relationship with the canyon through the 1100s and 1200s (John Stein, personal communication 1996). Furthermore, another study suggests that a north-south alignment between Chaco Canyon and Casas Grandes, a Mesoamerican site 630 km south of Chaco, developed in the 1300s, extended the earlier north-south axis from Aztec through Chaco (Lekson 1996).

28. Preliminary results of the Project's study of elevated horizons in the views to astronomy from certain major buildings suggest that the orientations of most of the interbuilding bearings to astronomical events on the sensible horizon (as shown in Table 9.3) are within 3° of the same astronomical events on the visible horizon. Exceptions to this generality appear to be the interbuilding bearings from Pueblo del Arroyo and Peñasco Blanco to Kin Bineola, from Kin Kletso to Pueblo Alto, and from Chetro Ketl to Kin Kletso.

29. In regard to the techniques used for orienting and interrelating buildings on astronomical bearings, the Solstice Project's experiments have shown that the cardinal directions can be determined with shadow and light to within one quarter of a degree (Solstice Project, prepublished report 1988; see also Williamson 1984:144). Recordings of the shadows cast by a vertical object onto a flat surface during several hours of the sun's midday passage indicate the cardinal directions. If this were done at a site with flat horizons toward the lunar standstills, at the time of the lunar standstills on that same surface where the cardinal directions would be recorded, the azimuths of the rising and setting standstill moons could also be recorded. It is possible that the Chacoan architects and planners used such a recording of the solar-lunar azimuths for incorporating lunar orientations in their buildings and in the interrelationships of their buildings, instead of waiting for the recurrence of the lunar events on the local horizons. The wait for the recurrence of the lunar major standstill would be 18 to 19 years. The Solstice Project has also shown that interrelating the buildings which are not intervisible could have been done with quite simple intersite surveying techniques.

30. See note 5.

31. It has been suggested that the Mayas' interest in the lunar eclipse cycle may have involved knowledge of the lunar standstill cycle (Dearborn 1992). Floyd Lounsbury (personal communication 1982) expressed a similar opinion a number of years ago.

32. W. J. Judge and J. M. Malville speculate on Chaco as a center for lunar eclipse prediction (1993), and the Malvilles suggest that ceremonial pilgrimage to Chaco Canyon was scheduled to the solar and lunar cycles (Malville and Malville 1995).

33. For ethnographic reports on the cosmology of the historic Pueblo Indians, see Sofaer, Marshall, and Sinclair (1989); Sofaer and Sinclair (1987); Sofaer, Sinclair, and Doggett (1982); Sofaer, Zinser, and Sinclair (1979); and Williamson (1984).

34. M. C. Stevenson (1894:143): "The moon is father to the dead as the sun is father to the living."

35. Fred Eggan's studies of the Hopi "roads" suggested to him several parallels with the Chacoans' use of roads. Eggan noted that at Hopi the spirits of the dead emerge from the world below and travel on symbolic roads to visit with the living, and that the Great North Road of Chaco appears to have been built to join the ceremonial center symbolically with the direction north and with the world below (Fred Eggan, personal communication 1990).

36. Alfonso Ortiz, *The Sun Dagger* documentary film (Solstice Project 1982).

REFERENCES

Ahlstrom, Richard Van Ness
1985. *Interpretation of Archeological Tree-Ring Dates.* Ph.D. dissertation, University of Arizona, Tucson.

Aveni, Anthony F.
1980. *Skywatchers of Ancient Mexico.* University of Texas Press, Austin.

Broda, Johanna
1982. "Astronomy, *Cosmovisión,* and Ideology in Prehispanic Mesoamerica," in *Ethnoastronomy and Archaeoastronomy in the American Tropics,* edited by Anthony F. Aveni and Gary Urton, pp. 81–110. Annals of the New York Academy of Sciences.

1993. "Archaeoastronomical Knowledge, Calendrics, and Sacred Geography in Ancient Mesoamerica," in *Astronomies and Culture,* edited by C. Ruggles and N. Saunders, pp. 253–295. University Press of Colorado, Niwot.

Clancy, Flora S.
1994. "Spatial Geometry and Logic in the Ancient Maya Mind. Part 1: Monuments," in *Seventh Palenque Round Table, 1989,* Merle Greene Robertson, gen. ed., Virginia M. Fields, vol. ed., pp. 237–242. Pre-Columbian Research Institute, San Francisco.

Cordell, Linda S.
1984. *Prehistory of the Southwest.* Academic Press, Orlando, Florida.

Crown, Patricia L., and W. James Judge, Jr.
1991. "Synthesis and Conclusions," in *Chaco and Hohokam: Prehistoric Regional Systems in the American Southwest,* edited by Patricia L. Crown and W. James Judge, Jr., pp. 293–308. School of American Research Press, Santa Fe.

Dearborn, David D.
1992. "To the Limits," *Archaeoastronomy and Ethnoastronomy News: Quarterly Bulletin of the Center for Archaeoastronomy* 3: 1, 4.

Ford, Dabney
1993. "The Spadefoot Toad Site: Investigations at 29SJ629," in *Marcia's Rincon and the Fajada Gap Pueblo II Community, Chaco Canyon, New Mexico,* edited by Thomas C. Windes, Vol. I, Appendix H. Reports of the Chaco Center 12. National Park Service, Albuquerque.

Fowler, Andrew P., and John Stein
1992. "The Anasazi Great House in Time, Space, and Paradigm," in *Anasazi Regional Organization and the Chaco System,* edited by David E. Doyel, pp. 101–122. Maxwell Museum of Anthropology, Anthropological Papers 5. University of New Mexico, Albuquerque.

Fritz, John M.
1978. "Paleopsychology Today: Ideational Systems and Human Adaptation in Prehistory," in *Social Archeology, Beyond Subsistence and Dating,* edited by Charles L. Redman *et al.,* pp. 37–59. Academic Press, New York.

Harrison, Peter D.
1994. "Spatial Geometry and Logic in the Ancient Mayan Mind. Part 2: Architecture," in *Seventh Palenque Round Table, 1989,* Merle Greene Robertson, gen. ed., Virginia M. Fields, vol. ed., pp. 243–252. Pre-Columbian Research Institute, San Francisco.

Hawkins, Gerald S.
1973. *Beyond Stonehenge.* Harper and Row, New York.

Judge, W. James, Jr.
1984. "New Light on Chaco Canyon," in *New Light on Chaco Canyon,* edited by David G. Noble, pp. 1–12. School of American Research Press, Santa Fe.
1989. "Chaco Canyon—San Juan Basin," in *Dynamics of Southwest Prehistory,* edited by Linda S. Cordell and George J. Gumerman, pp. 209–262. Smithsonian Institution Press, Washington, D.C.

Judge, W. James, Jr., and J. McKim Malville
2004. "Calendrical Knowledge and Ritual Power," in *Chimney Rock: The Ultimate Outlier,* edited by J. McKim Malville, pp. 151–163. Lexington Books, Lanham, Md.

Kincaid, Chris, editor
1983. *Chaco Roads Project Phase I: A Reappraisal of Prehistoric Roads in the San Juan Basin.* Bureau of Land Management, Albuquerque.

Lekson, Stephen H.
1984. *Great Pueblo Architecture of Chaco Canyon.* National Park Service, Albuquerque.
1991. "Settlement Patterns and the Chacoan Region," in *Chaco and Hohokam: Prehistoric Regional Systems in the American Southwest,* edited by P. L. Crown and W. J. Judge, Jr., pp. 31–56. School of American Research Press, Santa Fe.
1996. "Chaco and Casas." Paper presented at the 61st Annual Meeting of the Society for American Archeology.

Lekson, Stephen H., Thomas C. Windes, John R. Stein, and W. James Judge Jr.
1988. "The Chaco Canyon Community." *Scientific American* (July): 100–109.

Loose, Richard W., and Thomas R. Lyons
1977. "The Chetro Ketl Field: A Planned Water Control System in Chaco Canyon," in *Remote Sensing Experiments in Cultural Resource Studies,* assembled by Thomas R. Lyons, pp. 133–156. Reports of the Chaco Center 1, National Park Service, Washington, D.C.

McCluskey, Stephen C.
1977. "The Astronomy of the Hopi Indians." *Journal for the History of Astronomy* 8: 174–195.

Malville, J. McKim
2005. "Ancient Space and Time in the Canyons," in *Canyon Spirits: Beauty and Power in the Ancestral Puebloan World,* essays by S. H. Lekson and J. M. Malville, photographs by J. L. Ninnemann, foreword by F. C. Lister, pp. 65–86, University of New Mexico Press, Albuquerque.

1993. "Astronomy and Social Integration among the Anasazi," in *Proceedings of the Anasazi Symposium, 1991,* edited by Jack E. Smith and Art Hutchinson, pp. 155–166. Mesa Verde Museum Association.

Malville, J. McKim, and Nancy J. Malville
1995. "Pilgrimage and Astronomy at Chaco Canyon, New Mexico." Paper presented at the National Seminar on Pilgrimage, Tourism, and Conservation of Cultural Heritage, January 21–23, Allahabad, India.

Malville, J. McKim, and Claudia Putnam
1989. *Prehistoric Astronomy in the Southwest.* Johnson Books, Boulder.

Malville, J. McKim, Frank W. Eddy, and Carol Ambruster
1991. "Moonrise at Chimney Rock." *Journal for the History of Astronomy* (Supplement 1b: Archeoastronomy) 16: s34–s50.

Marshall, Michael P.
1997. "The Chacoan Roads—A Cosmological Interpretation," in *Anasazi Architecture and American Design,* edited by Baker H. Morrow and V. B. Price, pp. 62–74. University of New Mexico Press, Albuquerque.

Marshall, Michael P., and David E. Doyel
1981. *An Interim Report on Bis sa'ni Pueblo, with Notes on the Chacoan Regional System.* Manuscript on file, Navajo Nation Cultural Resource Management Program, Window Rock, Arizona.

Marshall, Michael P., John R. Stein, Richard W. Loose, and Judith E. Novotny
1979. *Anasazi Communities of the San Juan Basin.* Public Service Company of New Mexico, Albuquerque.

Morrow, Baker H., and V. B. Price, eds.
1997. *Anasazi Architecture and American Design.* University of New Mexico Press, Albuquerque.

Neitzel, Jill
1989. "The Chacoan Regional System: Interpreting the Evidence for Sociopolitical Complexity," in *The Sociopolitical Structure of Prehistoric Southwestern Societies,* edited by Steadman Upham, Kent G. Lightfoot, and Roberta A. Jewett, pp. 509–556. Westview Press, Boulder.

Powers, Robert P., William B. Gillespie, and Stephen H. Lekson
1983. *The Outlier Survey: A Regional View of Settlement in the San Juan Basin.* National Park Service, Albuquerque.

Reyman, Jonathan E.
1976. "Astronomy, Architecture, and Adaptation at Pueblo Bonito." *Science* 193: 957–962.

Roney, John R.
1992. "Prehistoric Roads and Regional Integration in the Chacoan System," in *Anasazi Regional Organization and the Chaco System,* edited by David E. Doyel, pp. 123–132. Maxwell Museum of Anthropology, Anthropological Papers 5. University of New Mexico, Albuquerque.

Sebastian, Lynne
1991. "Sociopolitical Complexity and the Chaco System," in *Chaco and Hohokam: Prehistoric Regional Systems in the American Southwest,* edited by Patricia L. Crown and W. James Judge, Jr., pp. 107–134. School of American Research Press, Santa Fe.
1992. "Chaco Canyon and the Anasazi Southwest: Changing Views of Sociopolitical Organization," in *Anasazi Regional Organization and the Chaco System,* edited by David E. Doyel, pp. 23–34. Maxwell Museum of Anthropology, Anthropological Papers 5. University of New Mexico, Albuquerque.

Sinclair, Rolf M., and Anna Sofaer
1993. "A Method for Determining Limits on the Accuracy of Naked-Eye Locations of Astronomical Events," in *Archaeoastronomy in the 1990s,* edited by Clive Ruggles, pp. 178–184. Group D Publications, Loughborough, U.K.

Sinclair, Rolf M., Anna Sofaer, John J. McCann, and John J. McCann, Jr.
1987. "Marking of Lunar Major Standstill at the Three-Slab Site on Fajada Butte." *Bulletin of the American Astronomical Society* 19:1043.

Sofaer, Anna
1994. "Chacoan Architecture: A Solar-Lunar Geometry," in *Time and Astronomy at the Meeting of Two Worlds,* edited by Stanislaw Iwaniszewski et al., pp. 265–278. Warsaw University, Poland.
1997. "The Primary Architecture of the Chacoan Culture: A Cosmological Expression," in *Anasazi Architecture and American Design,* edited by Baker H. Morrow and V. B. Price, pp. 88–130. University of New Mexico Press, Albuquerque.
2006. "Pueblo Bonito Petroglyph on Fajada Butte: Solar Aspects," in *Celestial Seasonings: Connotations of Rock Art, 1994 IRAC Proceedings,* Rock Art-World Heritage, edited by E. C. Krupp, pp. 397–402. American Rock Art Research Association, Phoenix.

Sofaer, Anna, and Rolf M. Sinclair
1987. "Astronomical Markings at Three Sites on Fajada Butte," in *Astronomy and Ceremony in the Prehistoric Southwest,* edited by John Carlson and W. James Judge Jr., pp. 43–70. Maxwell Museum of Anthropology, Anthropological Papers No. 2. University of New Mexico, Albuquerque.
1989. "An Interpretation of a Unique Petroglyph in Chaco Canyon, New Mexico," in *World Archaeoastronomy,* edited by Anthony F. Aveni, p. 499. Cambridge University Press, Cambridge, U.K.

Sofaer, Anna, Michael P. Marshall, and Rolf M. Sinclair
1989. "The Great North Road: A Cosmographic Expression of the Chaco Culture of New Mexico," in *World Archaeoastronomy,* edited by Anthony F. Aveni, pp. 365–376. Cambridge University Press, Cambridge, U.K.

Sofaer, Anna, Rolf M. Sinclair, and LeRoy Doggett
1982. "Lunar Markings on Fajada Butte, Chaco Canyon, New Mexico," in *Archaeoastronomy in the New World,* edited by Anthony F. Aveni, pp. 169–181. Cambridge University Press, Cambridge, U.K.

Sofaer, Anna, Rolf M. Sinclair, and Joey B. Donahue
1991. "Solar and Lunar Orientations of the Major Architecture of the Chaco Culture of New Mexico," in *Colloquio Internazionale Archeologia e Astronomia,* edited by G. Romano and G. Traversari, pp. 137–150. Rivista di Archaeologia, Supplementi 9, edited by Giorgio Bretschneider. Rome, Italy.

Sofaer Anna, Rolf M. Sinclair, and Reid Williams
1987. "A Regional Pattern in the Architecture of the Chaco Culture of New Mexico and its Astronomical Implications." *Bulletin of the American Astronomical Society* 19: 1044.

Sofaer, Anna, Volker Zinser, and Rolf M. Sinclair
1979. "A Unique Solar Marking Construct." *Science* 206: 283–291.

Stein, John R.
1989. "The Chaco Roads—Clues to an Ancient Riddle?" *El Palacio* 94: 4–16.

Stein, John R., and Stephen H. Lekson
1992. "Anasazi Ritual Landscapes," in *Anasazi Regional Organization and the Chaco System,* edited by David E. Doyel, pp. 87–100. Maxwell Museum of Anthropology, Anthropological Papers 5. University of New Mexico, Albuquerque.

Stein, John R., and Peter J. McKenna
1988. *An Archaeological Reconnaissance of a Late Bonito Phase Occupation near Aztec Ruins National Monument.* National Park Service, Southwest Cultural Resource Center, Santa Fe.

Stein, John R., Judith E. Suiter, and Dabney Ford
1997. "High Noon in Old Bonito: Sun, Shadow, and the Geometry of the Chaco Complex," in *Anasazi Architecture and American Design*, edited by Baker H. Morrow and V. B. Price, pp. 133–148. University of New Mexico Press, Albuquerque.

Stevenson, Mathilda Coxe
1894. "The Sia," in *Eleventh Annual Report of the Bureau of Ethnology.* Smithsonian Institution, Washington, D.C.

Tedlock, Barbara
1983. "Zuñi Sacred Theater." *American Indian Quarterly* 7:93–109.

Toll, Henry Walcott
1991. "Material Distributions and Exchange in the Chaco System," in *Chaco and Hohokam: Prehistoric Regional Systems in the American Southwest*, edited by Patricia L. Crown and W. James Judge, Jr., pp. 77–108. School of American Research Press, Santa Fe.

Vivian, R. Gwinn
1990. *The Chacoan Prehistory of the San Juan Basin.* Academic Press, San Diego.

Williamson, Ray A.
1984. *Living the Sky.* Houghton Mifflin, Boston.

Williamson, Ray A., Howard J. Fisher, and Donnel O'Flynn
1977. "Anasazi Solar Observatories," in *Native American Astronomy*, edited by Anthony F. Aveni, pp. 203–218. University of Texas Press, Austin.

Williamson, Ray A., Howard J. Fisher, Abigail F. Williamson, and Clarion Cochran
1975. "The Astronomical Record in Chaco Canyon, New Mexico," in *Archaeoastronomy in Pre- Columbian America*, edited by Anthony F. Aveni, pp. 33–43. University of Texas Press, Austin.

Windes, Thomas C.
1978. *Stone Circles of Chaco Canyon, Northwestern New Mexico.* Reports of the Chaco Center No. 5, Division of Chaco Research, National Park Service, Albuquerque.

1987. *Investigations at the Pueblo Alto Complex, Chaco Canyon, New Mexico, 1975–1979. Vol. 1: Summary of Test and Excavations at the Pueblo Alto Community.* Publications in Archeology 18F. National Park Service, Santa Fe.

Zeilik, Michael
1984. "Summer Solstice at Casa Rinconada: Calendar, Hierophany, or Nothing?" *Archaeoastronomy* 7: 76–81.

6

Chacoan Architecture: A Solar-Lunar Geometry

Summary: *We find that certain internal angles of the fourteen major buildings of the ancient Chacoan culture of New Mexico correlate with the angles between the azimuths of the lunar standstills and the cardinal directions. We also find that this solar-lunar geometry internal to the major buildings is integrated with the external solar-lunar orientations and the solar-lunar regional pattern of these buildings. Our earlier analysis had found that the orientations of twelve of the fourteen major Chacoan buildings are to the rising and setting azimuths of the sun and moon at the extremes and midpositions of their cycles: i.e. the cardinals, the solstices and the lunar standstills (Sofaer 1993). This earlier work also showed that the Chacoans coordinated these solar-lunar orientations with the locations of their major buildings to form a solar-lunar regional pattern that is symmetrically organized about the ceremonial center of Chaco Canyon (Sofaer 1993).*

The reader is referred to several previous papers for details of earlier Solstice Project studies and for references to archaeological background on the Chacoan culture.

Anna Sofaer

Published in
Time and Astronomy
at the Meeting of Two Worlds,
edited by
Stanislaw Iwaniszewski, et al.
(Warsaw University,
Warsaw, Poland, 1994).
Originally presented at the
International Symposium,
Frombork, Poland, 1992.
Organized by the Department
of Historical Anthropology,
Institute of Archaeology,
Warsaw University.

FROM APPROXIMATELY A.D. 900 TO 1130, the Chacoan culture, a prehistoric Pueblo society, constructed numerous multi-storied buildings and extensive roads throughout the 80,000 square kilometers of the arid San Juan Basin of northwestern New Mexico. Recent evidence suggests that the culture extended over an area two to four times that of the San Juan Basin (Lekson *et al.* 1988). Chaco Canyon, where most of the largest buildings were constructed, was the center of the culture. The canyon is located close to the center of the high desert of the San Juan Basin.

The Chacoan buildings were constructed on a massive scale, with large public plazas and elaborate road connections (Lekson *et al.* 1988; Lekson 1984; Marshall *et al.* 1979). The buildings in our study are the fourteen largest (by room count) in the Chaco cultural region (Powers *et al.* 1983). They contained 115-695 rooms, as well as numerous kivas, including great kivas, the large ceremonial chambers of prehistoric Pueblo culture, and they were up to four stories high. The extensive planning and engineering involved in the multi-storied buildings is evident in their symmetric and grid-like layout. Ten of the buildings are located in Chaco Canyon, and four are located outside the canyon.

Utilitarian considerations appear not to have significantly dominated or constrained the Chacoans' choice of specific locations for the major buildings (Sofaer 1993). Archaeological investigations indicate that major buildings in Chaco Canyon were not built or used primarily for household occupation (Lekson *et al.* 1988). Evidence of periodic large-scale ceremonial breakage of vessels at key central buildings indicates, however, that Chaco Canyon and certain of the major buildings were used for seasonal ceremonial visitations by great numbers of residents of the outlying communities (Lekson *et al.* 1988).

Scholars have commented extensively on the impractical and enigmatic aspects of the Chacoan buildings, describing them as "overbuilt and overembellished" and proposing that they were built primarily for public image and ritual expression (Sofaer 1993; Lekson *et al.* 1988). It has recently been suggested that the Chacoan buildings were developed as expressions of the Chacoans' "concepts of the cosmos" (Stein and Lekson 1990). This report further proposes that "Chaco and its hinterland are related by a canon of shared design concepts" and that the Chacoan architecture is a "common ideational bond" across a "broad geographic space."

The Chacoans also constructed over 200 km of roads which were of great width (averaging 9 meters) and unusual linearity over long distances. Recent evidence suggests that certain of these roads, which were clearly overbuilt if they were intended to serve purely utilitarian purposes, may have been constructed as cosmographic expressions (Sofaer *et al.* 1989).

SOLAR-LUNAR GEOMETRY INTERNAL TO THE MAJOR CHACOAN BUILDINGS

Our recent analysis of the geometry internal to the major Chacoan buildings shows that certain primary internal angles of the fourteen major buildings cluster in two groups of angles and that these groups of angles appear to express the relationship between the solar and lunar cycles. (Figure 1 illustrates, for the Chaco region, the rising and setting azimuths of the suns at the summer and winter solstices and of the moons at the major and minor standstills.)

We refer to the angles of interest in our study as the "primary internal solar-lunar angles." These angles are formed by the hypotenuse and longer leg of the triangles we find in the internal geometry of each of the buildings. The two groups of primary internal solar-lunar angles used in the major Chacoan buildings are 22°-27° (Table I and Figure 2 top) and 33°-39° (Table I and Figure 3 top). The angle between the cardinal east-west direction and the lunar minor standstill azimuth is 22.9° (Figure 2 bottom); and the angle between the east-west direction and the lunar major standstill is 35.7° (Figure 3 bottom). The primary internal solar-lunar angles help determine the shapes and proportions of the major Chacoan buildings.

In thirteen of the fourteen buildings, the hypotenuse of the primary internal solar-lunar angle is the theoretical line formed by a diagonal between two external corners of primary walls (eleven instances) or a chord between one external corner of a primary wall and a point on the external surface of a curved back wall (two instances).[1] The "primary walls" are the principal long back walls, the side walls, and the curved back walls. In eleven of the fourteen buildings, the longer leg of the primary internal solar-lunar angle is the line formed by a primary wall (ten instances) or a theoretical chord between the external corners of a curved primary wall (one instance). [2,3]

The frequency of the occurrence of primary internal solar-lunar angles in the fourteen major buildings suggests that these angles are a fundamental and characteristic design feature of the

Chacoan architecture. These angles occur at least twenty-six times in the fourteen buildings.

The primary internal solar-lunar angles of nine of the fourteen major Chacoan buildings are between 22 and 27°; and the primary internal solar-lunar angles of eight of these same fourteen major Chacoan buildings are between 33 and 39°. It is interesting to note that three buildings (Aztec, Kin Bineola and Tsin Kletzin) incorporate angles from both groups of primary internal solar-lunar angles.

We also found that there are solar and lunar orientations in the diagonals of eight buildings (eleven instances) and in the line between the external corners of one building (Una Vida)[4] (Table I and Figures 2 and 3). All of these orientations are within 3.1° of the respective cardinal and lunar azimuths; and six are within 1.5° of these azimuths. (This range of accuracy is consistent with the range of accuracy exhibited in other aspects of Chacoan archaeoastronomical constructions, Sofaer 1993.)

In four of the buildings, the orientations of the diagonals which are to the solar or lunar azimuths are also on the bearings to other buildings (five instances).[5] With the exception of Wijiji's solstitial relationship to the center, these diagonal lunar inter-building relationships lie on the bearings of the astronomical inter-building relationships, which we describe in the next section.

(It is of interest that the angle of the sun's rays could be seen as the hypotenuse of a right angled triangle formed by a vertical shadow casting object, such as a building wall, and a flat surface. At solar meridian passage on equinox day, a time that is the middle of the day and of the year, the angle between the hypotenuse and the long (vertical) leg of this triangle, at the latitude of Chaco, would be 36°. In addition, it is also of interest that the two groups of internal solar-lunar angles are close to those in the right-angled triangles defined by the two simplest Pythagorean triplets: the triplet (3, 4, 5) generates the angle 36.9° and the triplet (5, 12, 13) generates 22.6°. We are investigating the role these triplets may have played in the aesthetic and other considerations of Chacoan construction.)

We have not attempted in the short compass of this paper to deal exhaustively with all internal solar-lunar angles. We note, however, that, in the case of Kin Bineola for example, it contains, in addition to its primary solar-lunar angles as described above, other secondary internal solar-lunar angles (Figure 3). These secondary internal angles are formed by extending the lines from the eastern and western sides of the central room block to points on the external surface of the long back wall and diagonals from those points to the external corners of the western and eastern walls, respectively.

The primary internal solar-lunar angles of nine buildings were obtained through our survey of the orientations and lengths of the principal and short walls. (The results and methodology of this survey were previously reported, Sofaer 1993.) The primary internal solar-lunar angles of four buildings and the secondary internal solar-lunar angles of Kin Bineola were obtained by taking angles off maps of these buildings (Lekson 1984; Marshall *et al.* 1979).

The orientations of the diagonals of the seven rectangular buildings, in which the diagonals are related to solar or lunar azimuths, were obtained through surveying. The orientations of the line between the external corners of Una Vida and of the secondary diagonals in Kin Bineola were obtained by relating the angles taken off maps of these buildings (Lekson 1984; Marshall *et al.* 1979) to our survey data on the orientations of the principal walls.

SIMILAR ASTRONOMICAL EXTERNAL EXPRESSIONS IN THE MAJOR CHACOAN BUILDINGS

We find that the internal geometry of the major Chacoan buildings displays the same interests in the solar and lunar cycles that, in our earlier study, we had found expressed in the buildings' external orientations and inter-building relationships.

Our earlier survey and analysis had found that the orientations of the principal (primary, long back) walls, perpendiculars (to the principal walls) and major axes of twelve of the fourteen major Chacoan

Solar-Lunar Orientations

Buildings	Primary (Secondary) Internal Angles		Internal Geometry	Principal Wall or Axis	Perpendicular
Hungo Pavi	23.6°	24.1°	-66.3°: lunar minor 66.0°: lunar minor	-89.9°: cardinal	
Chetro Ketl	24.0°			69.7°: lunar minor	
Aztec I	24.1°	25.3°		62.5°: solstice	
Kin Kietso	27.0°	26.5°	-87.8°: cardinal	-65.7°: lunar minor	
Pueblo Alto	22.7°	23.9°	-67.3°: lunar minor 65.1°: lunar minor	88.3°: cardinal	
Tsin Kletzin	25.0°		-64.1° ± 1°: lunar minor	89° ± 1°: cardinal	
Salmon Ruin	22.6°	22.5°	88.4°: cardinal	65.8°: lunar minor	
Pueblo del Arroyo	25.1°	26.4°	-1.6°: cardinal°		-65.3°: lunar minor
Kin Bineola	23.7°	24.5°	54.5° ± 1° lunar major		
Aztec II	36.1°	37.7°		62.5°: solstice	
Pueblo Pintado	38.5°			70.2°: lunar minor	
(Kin Bineola)	(34.5°)	(34.3°)	(-66.7° ± 1°: lunar minor)		
Wijiji	34.5°	34.3°	-62.0°: solstice		
Tsin Kletzin		38.5°		89° ± 1°: cardinal	
Penasco Blanco	35.5°	33.0°			-53.9°: lunar major
Pueblo Bonito	33.2°			0.26°: cardinal	-89.7°: cardinal
Una Vida	35.2°		-88.8° ± 1°: cardinal		54.8°: lunar major

© Solstice Project 1992

TABLE 1.   *Two groups of primary internal solar-lunar angles used in the Chacoan architecture.*

118

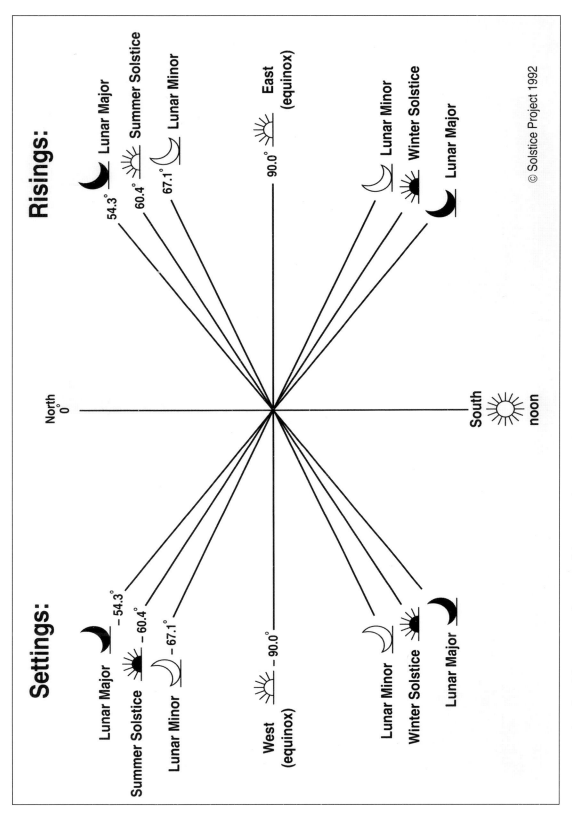

FIGURE 1. *Azimuths of the rising and setting of the sun and moon at the extremes and mid-positions of their cycles, at the latitude (36° North) of Chaco Canyon. The meridian passage is also indicated. The lunar extremes are the northern and southern limits of the moon's rising and setting at the major and minor standstills. (Suzanne Young, By Design Graphics, Copyright © 1992 Solstice Project.)*

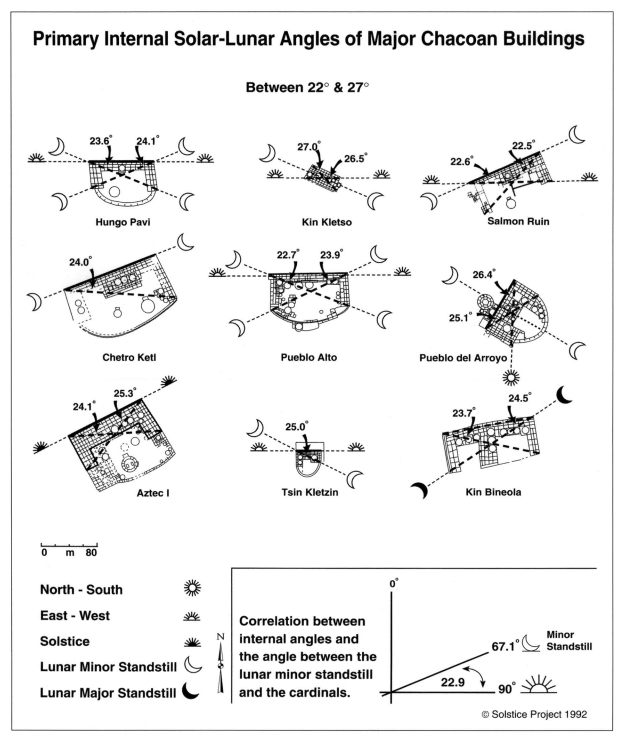

FIGURE 2. *Primary internal solar-lunar angles of nine major Chacoan buildings between 22° and 27° (top); correlation between the buildings' internal angles and the angle between the lunar minor standstill and the cardinal directions (bottom). The diagram also shows the solar-lunar orientations of the buildings. (Suzanne Young, By Design Graphics, Copyright © 1992 Solstice Project.)*

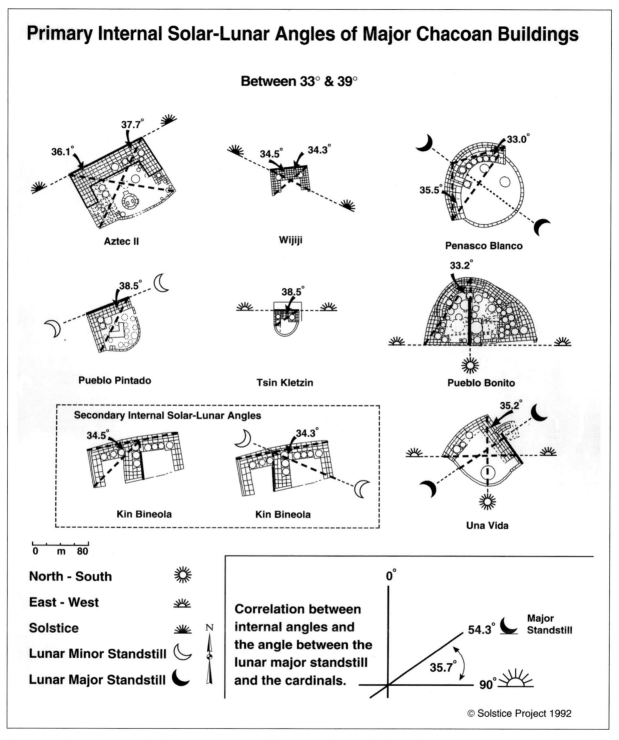

FIGURE 3.   *Primary internal solar-lunar angles of eight major Chacoan buildings between 33° and 39° (top); correlation between the buildings' internal angles and the angle between the lunar major standstill and the cardinal directions (bottom). The diagram also shows the solar-lunar orientations of the buildings. (Suzanne Young, By Design Graphics, Copyright © 1992 Solstice Project.)*

buildings are to one of the four solar and lunar az-imuths (Sofaer 1993)[6] (Table I and Figures 2 and 3).

The results of our recent study of the internal geometry showing astronomical orientations in the diagonals of eight buildings and the mid-axis of one building, when related to these earlier find-ings of astronomical orientations in the principal walls, perpendiculars, and mid-axes of the major Chacoan buildings, show that seven buildings have orientations to both the sun and moon. That is to say, seven buildings have, in addition to the solar or lunar orientations of their principal walls or perpendiculars, orientations on their diagonals or major axes to the sun or moon.[7]

We note that the solar-lunar geometry internal to the major Chacoan buildings is also evident in the symmetric organization of the Chacoan astro-nomical regional pattern.

Our earlier analysis showed that numerous bearings between thirteen of the fourteen major buildings align with the azimuths of the solar and lunar phenomena associated with the individual buildings (Sofaer 1993). (Only one major building, Salmon Ruin, is not related in this manner to another building.) These astronomical inter-build-ing relationships form a regional pattern that is symmetrically organized about the central archi-tectural complex of Chaco Canyon. The pattern encompasses approximately 5,000 km.[2]

The symmetric pattern of astronomical inter-building relationships is organized by the same angles that form the internal geometry of the buildings. The central buildings are organized by cardinal inter-building relationships; and the outlying major buildings are related to these cen-tral buildings on the bearings to the standstill moons, which bearings are symmetric about the cardinal axes of the central complex. Thus, the two angles that are internal to the Chacoan astronomical regional pattern are—as in the case of the angles internal to the individual build-ings—approximately the same as (1) the angle between the east-west direction and the lunar minor standstill azimuth (22.9°) and (2) the angle between the east-west direction and the lunar major standstill (35.7°).

PUEBLO ALTO

Our analysis suggests that there are fundamental principles on which the Chacoan buildings are founded. These principles govern the design, ori-entation, and location of the major buildings. They can be illustrated by examining in detail the astro-nomical relationships in one central building, Pueblo Alto.

We see in Pueblo Alto both the order and complexity of the Chacoans' multi-layered archi-tectural expression of solar-lunar astronomy and geometry. As Figure 4 shows, there is a unifying symmetry, based on the solar and lunar patterns, that interrelates Pueblo Alto's solar-lunar internal geometry, external orientations and inter-building relationships.

First, Pueblo Alto's primary internal angles, i.e. the angles between its primary, long back wall and its diagonals, are 22.7° and 23.9°, which correlate closely with the angle between the minor standstill moon's azimuth and the east-west direction (22.9°). Second, it has lunar orientations in its diagonals. Third, it is related along the orientations of its diag-onals to the outlying buildings Kin Bineola and Pueblo Pintado, 18.1 km and 26.7 km distant, on bearings that are to the rising and setting of the southern minor standstill moon and that are sym-metric about the north-south axis of the central complex. Fourth, it has cardinal orientations in its primary, long back wall and in its perpendicular. And fifth, it is related on its north-south perpendi-cular orientation to Tsin Kletzin, 3.7 km distant, in an alignment that forms the axis of symmetry of the central complex of buildings and of the rela-tionships of most of the outlying major buildings to the central complex. And on this same north-south axis, Pueblo Alto is related to Aztec, a major Chacoan building which is located 86 km to the north of Pueblo Alto.

Thus, Pueblo Alto's internal and external rela-tionships are defined by mid-points in the daily and yearly solar cycles and by extreme positions in the 18.6 year lunar standstill cycle. At equinox, the sun rises and sets along Pueblo Alto's east-west wall; and, on each day when the sun passes the

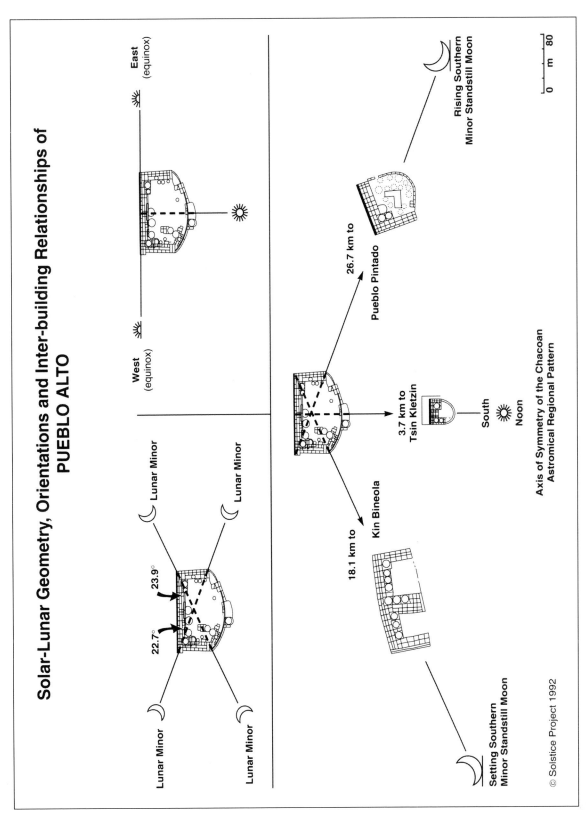

FIGURE 4. *Solar-lunar geometry, orientations and inter-building relationships of Pueblo Alto. (Suzanne Young, By Design Graphics, Copyright © 1992 Solstice Project.)*

meridian, it is in alignment with the building's perpendicular axis. Pueblo Alto's major axes integrate the mid-passage of the sun's yearly cycle with the mid-passage of the sun's daily cycle. And furthermore, each day at solar meridian passage, the sun is also in alignment with the bearing between Pueblo Alto and Tsin Kletzin, the bearing which forms the north-south axis of symmetry of the Chacoan astronomical regional pattern. Finally, when the moon reaches its minor standstill, the northern and southern limits of its risings and settings will be on the azimuths that are the orientations of the diagonals of Pueblo Alto. And the minor standstill moon's rising and setting on these azimuths marks the bearings from Pueblo Alto and the central complex to Pueblo Pintado and Kin Bineola, as well as to other major Chacoan buildings.

CONCLUSION

Like Pueblo Alto, each of the other major Chacoan buildings has its own specific astronomical expressions which also connect it with the other major Chacoan buildings by common solar-lunar geometry, external orientations, and inter-building relationships. Each building is an individual expression of a common solar-lunar cosmology.

Other expressions of the Chacoans' interest in the solar and lunar cycles have been documented in light markings on petroglyphs (Sofaer *et al.* 1979, 1982). Solar and lunar orientations and inter-building relationships have also been reported in other, outlying Chaco-related buildings (Malville 1991).

A later paper will discuss the development over time of the solar-lunar design, orientation, and interrelationships of the major Chacoan buildings and their cultural implications.

There are numerous parallels to the Chacoan's architectural expression of cosmology among the cultures of Mesoamerica. It is known that the Chacoans had some cultural associations with Mesoamerica. The orientations, interrelationships and geometry of architectural complexes in Mesoamerica are frequently ordered by astronomical phenomena (Aveni 1980).

In the cosmology of historic Pueblo peoples, the sun and moon are of primary importance and their integration is constantly sought in the scheduling of ceremonies and the ritual traditions. The joining of the cardinal and solstice directions with the nadir and the zenith frequently defines, in Pueblo ceremony and myth, the sacred center, around which the recurring solar and lunar cycles revolve. Chaco Canyon and its architectural complex appear to have been such a center place.

ACKNOWLEDGMENTS

Several individuals made invaluable contributions to this paper. We especially appreciate Rolf M. Sinclair's (National Science Foundation) thoughtful analysis of the data on the Chacoan architecture, as well as his rigorous evaluation of naked-eye astronomical observations. John Stein (Navajo Nation) shared with us his unique insights into the geometry, as well as the archaeological record of the Chacoan buildings. We benefited from many helpful discussions regarding Chacoan archaeology and archaeoastronomy, and Pueblo cosmology, with Michael Marshall (Cibola Research Consultants), Stephen Lekson (Museum of New Mexico), John Roney (Bureau of Land Management), Dabney Ford (National Park Service), Philip Johnson (Ohio State University), Fred Eggan (Santa Fe), and Jay Miller (Seattle). David Dearborn (Lawrence Livermore National Laboratory), J. McKim Malville (University of Colorado) and LeRoy Doggett (U.S. Naval Observatory) gave us thoughtful reviews of this material. We again thank William Byler (Washington, D.C.) for his generous and astute editing of the final manuscript. The graphics were prepared by Suzanne Young (By Design Graphics). We appreciate the National Park Service's invaluable cooperation with our numerous field trips for the collection of data on the Chacoan buildings.

[Author's note: Subsequent to the presentation of this paper, the author discovered that one of the buildings in the sample of this study, Hungo Pavi, is oriented to −85.2°, rather than to

−89.9°, as reported at Frombork. Accordingly, the reader should be aware when reading the text and examining the diagram and tables. In the author's view, this new data does not affect the essential thesis of this paper.]

ENDNOTES

1. In the one building that does not conform to this pattern (Una Vida), the hypotenuse is formed by a primary wall.

2. In the three buildings that do not conform to this pattern, the longer leg is formed by the theoretical line (chord) between two external corners of a curved primary wall (Peñasco Blanco) or a perpendicular to the theoretical line between two external corners of a curved primary wall (Pueblo Bonito and Una Vida). In the case of Pueblo Bonito, the longer leg is the perpendicular formed by the primary central wall and its theoretical extension to a point on the external surface of the long curved primary wall. In the case of Una Vida, the longer leg is the theoretical perpendicular erected on the theoretical line between two external corners of the primary walls to the external corner of the other end of one of these primary walls.

3. The only building that departs from general rules for the formation of both the hypotenuse (diagonals and chords) and the longer leg (primary walls) of the primary internal solar-lunar angles is Una Vida. Yet, despite its exceptional character, Una Vida fits within the overall pattern of the buildings by virtue of the fact that it embodies a primary internal angle which is formed by a hypotenuse and longer leg and correlates with the angle between the major lunar standstill and the east-west cardinal direction.

4. The orientations of a diagonal of Salmon Ruin, Kin Kletso, and Pueblo del Arroyo, and of the line between the external corners of Una Vida, are to the cardinal directions; the orientation of a diagonal of Wijiji is to the solstice azimuths; the orientations of both diagonals of Hungo Pavi and Pueblo Alto, and one of the diagonals of Tsin Kletzin, and one of the secondary diagonals of Kin Bineola, are to the lunar minor standstill azimuths; and the orientation of a diagonal of Kin Bineola is to the lunar major standstill azimuths.

5. The diagonal of Kin Bineola is on the bearing to Peñasco Blanco, as well as to the lunar major standstill azimuth; the diagonals of Hungo Pavi and Wijiji are on the bearings to the central complex of buildings, as well as to the lunar minor standstill azimuth and the solstice azimuth, respectively; the diagonals of Pueblo Alto are on the bearings to the outlying buildings, Pueblo Pintado and Kin Bineola, as well as to the lunar minor standstill azimuths.

6. The orientations of the principal walls or mid-axis of Pueblo Bonito, Hungo Pavi, Pueblo Alto, and Tsin Kletzin are to the cardinal directions; and the orientations of the principal walls or perpendiculars of Chetro Ketl, Kin Kletso, Pueblo del Arroyo, Pueblo Pintado and Salmon Ruin are to the lunar minor standstill azimuths; and the orientations of the mid-axis or perpendicular of Peñasco Blanco and Una Vida are to the lunar major standstill azimuths; and the principal wall of Aztec is to the solstice azimuths. (There appears to be no significant astronomical association for the orientations of the principal walls or perpendiculars of two buildings, Wijiji and Kin Bineola.)

7. Hungo Pavi, Pueblo Alto, Salmon Ruin, Kin Kletso, Pueblo del Arroyo, and Tsin Kletzin have orientations to the cardinal directions and the lunar minor standstill azimuths; and Una Vida has orientations to the cardinal directions and the lunar major standstill azimuths. (Kin Bineola has orientations on its primary diagonal to the major standstill moon and on its secondary diagonal to the minor standstill moon.)

REFERENCES

Aveni, Anthony F.
   1980. *Skywatchers of Ancient Mexico*. University of Texas Press, Austin.

Lekson, Stephen H.
   1984. *Great Pueblo Architecture of Chaco Canyon*. National Park Service, Albuquerque.

Lekson, Stephen H., Thomas C. Windes, John R. Stein, W. James Judge, Jr.
   1988. "The Chaco Canyon Community." *Scientific American* July, 1988: 100-109.

Malville, J. McKim
   1991. "Prehistoric Astronomy in the American Southwest." *Astronomy Quarterly* 8: 1-36.

Marshall, Michael P., John R. Stein, Richard W. Loose, Judith E. Novotny
1979. *Anasazi Communities of the San Juan Basin.* Public Service Company of New Mexico, Albuquerque.

Powers, Robert P., William B. Gillespie, Stephen H. Lekson
1983. *The Outlier Survey, A Regional View of Settlement in the San Juan Basin.* National Park Service, Albuquerque.

Sofaer, Anna, Volker Zinser, Rolf M. Sinclair
1979. "A Unique Solar Marking Construct," *Science* 206: 283-291.

Sofaer, Anna, Rolf M. Sinclair, Leroy Doggett
1982. "Lunar Markings on Fajada Butte, Chaco Canyon, New Mexico," in A. F. Aveni, editor *Archaeoastronomy in the New World,* pp. 169-181. Cambridge University Press, Cambridge.

Sofaer, Anna, Michael P. Marshall, Rolf M. Sinclair
1989. "The Great North Road: A Cosmographic Expression of the Chaco Culture of New Mexico," in A. F. Aveni, editor, *World Archaeoastronomy,* pp. 365-376. Cambridge University Press, Cambridge.

Sofaer, Anna
1993. "The Primary Architecture of the Chacoan Culture: An Integrated Cosmological Expression." Paper presented at Symposium on Anasazi Architecture and American Design, Mesa Verde, May 1991; later published in *Anasazi Architecture and American Design,* University of New Mexico Press.

Stein, John, and Stephen H. Lekson
1990. "Anasazi Ritual Landscapes." Paper presented at the 55th Annual Meeting, Society for American Archaeology, Las Vegas (1990). To be published by the Maxwell Museum of Anthropology.

The
Great
North
Road

7

The Great North Road:

A Cosmographic Expression of the Chaco Culture of New Mexico

This paper originally appeared in World Archaeoastronomy, *A. F. Aveni, editor, published by Cambridge University Press in 1989. Michael P. Marshall is associated with Cibola Research Consultants, Corrales, New Mexico. Rolf M. Sinclair was on the staff of the National Science Foundation, Washington, D.C.*

Anna Sofaer,
Michael P. Marshall
and
Rolf M. Sinclair

THE GREAT NORTH ROAD is one of the most enigmatic constructs of the ancient Chaco culture of New Mexico. Efforts to establish strict utilitarian purposes for its construction do not explain certain unique features of it. We suggest it is a cosmographic expression of the Chaco culture.

The Chaco society, a prehistoric Pueblo culture, flourished between A.D. 950 and 1150 throughout the 80,000 kilometers of the San Juan Basin of northwestern New Mexico (Marshall *et al.* 1979; Powers, Gillespie and Lekson, 1983; Cordell, 1984; Marshall and Sofaer, 1988) (Figure 29.1). Chaco Canyon was the center of this culture. Here the Chaco people constructed multi-storied buildings containing 100 to 700 rooms (Lekson, 1984). These structures are noted for their planned, symmetric organization, massive core-veneer masonry construction, and numerous great kivas, the large ceremonial chambers of the prehistoric Pueblo culture.

Descendants of the prehistoric Pueblo culture live today in the pueblos of New Mexico and Arizona. Ethnographic reports on the traditions of the historic Pueblo Indians suggest parallels between the historic and prehistoric and may provide insights into the general cosmological concepts of the prehistoric Chaco culture.

Astronomy played an important role in the Chaco culture. This is expressed in the cardinal alignments of the major axes of several large ceremonial structures at or near the center of the canyon (Williamson, Fisher and O'Flynn, 1977; Sofaer and Sinclair, 1986a), and in a complex set of solar and lunar markings on Fajada Butte, at the south entrance of

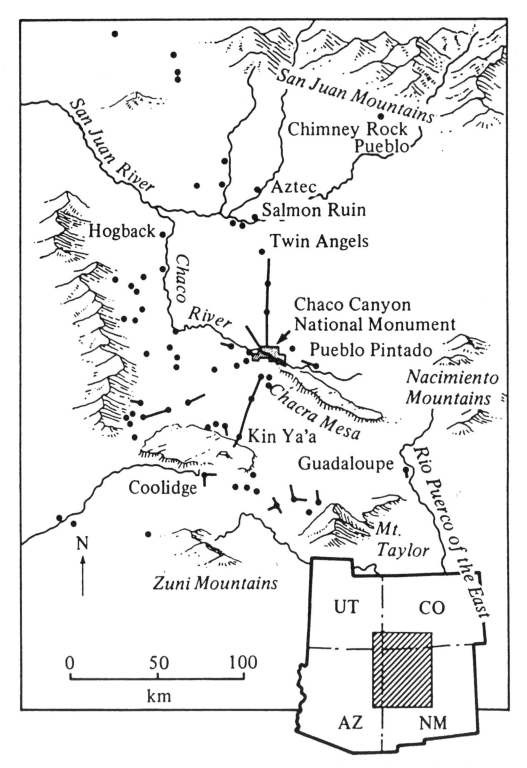

FIGURE 29.1. *Map of the San Juan Basin, showing major Chacoan sites and roads. The inset shows this region within present State boundaries. —— prehistoric roads; • outlying Chacoan communities. (Copyright © 1986 Carol Cooperrider; additional data supplied by John Roney.)*

FIGURE 29.2. *Portion of road on mesa above Pueblo Bonito. This unusually well-preserved section was formed by clearing to bedrock. (Copyright © 1985 David L. Brill.)*

the canyon (Sofaer, Zinser and Sinclair, 1979; Sofaer, Sinclair and Doggett, 1982; Sofaer and Sinclair, 1986a).

THE ROADS

Roads also played an important role in the Chaco culture, judging from their extent and the effort required for their design and construction (Kincaid, 1983). In the late florescence of the Chaco civilization (c. A.D. 1050 to 1125) elaborate, formalized roads were constructed (Figure 29.2). No archetype for these roads appears to have existed in the region before their development by the Chacoans, and a recent inventory shows they were not used after the Chaco civilization's peak, about A.D. 1140. The Chaco roads have generally been interpreted as arteries connecting communities for trade, transportation of goods and materials, and movement of population. These explanations of the roads' functions have been premised on a model of the Chaco culture that has envisioned Chaco Canyon as the political and economic center of a widespread trade and redistribution system extending throughout the San Juan Basin. (Vivian, 1983, provides a detailed summary of the various economic models which have been applied to the Chaco culture.)

The extensive religious architecture in Chaco Canyon suggests that the canyon may have served primarily as a ceremonial nexus for the outlying communities. Factors supporting this concept include: evidence from the middens of periodic intensive consumption of food at the large public structures (Judge, 1984); the dearth of burials and the presence of a few "high status" burials (Akins and Schelberg, 1984); and possible large-scale ceremonial breakage of ceramic vessels (Toll, 1984).

Consistent with this view of the religious function of Chaco Canyon, it appears that one of the Chaco roads—the Great North Road—and perhaps others, express religious considerations.

The Chaco roads have been noted for their great width and unusual linearity, and they have been described as "extensively engineered" (Nials, 1983, p. 6.26). The roads were developed by excavation to a smooth, level surface, and some included masonry construction.

Approximately 300 km of roads, including the Great North Road, have been documented in the last 15 years by aerial photography and ground investigation in numerous intensive studies. A further archaeological investigation of the Great North Road was recently conducted by the authors. This involved the inspection of all structural sites and many kilometers of roadway. Numerous sites were mapped and sampled, and a technical report concerning this work is in preparation (Marshall and Sofaer, 1988).

THE GREAT NORTH ROAD—DESCRIPTION

The Great North Road (Figure 29.3) has its origin in several routes which ascend by staircases carved into the cliff from Pueblo Bonito and Chetro Ketl in Chaco Canyon, which are the two largest structures of the Chaco region. These routes converge at Pueblo Alto, a large structure located close to the north rim of the canyon. From there the road runs 13° to the east of north for 3 km to Escavada Wash. It then heads within ½° of true north for 16 km, where it articulates with Pierre's Complex, an unusual cluster of small buildings on knobs and pinnacles. The road then heads close to 2° east of north for 31 km and ends at Kutz Canyon. It appears to terminate at three small, isolated sites, and a stairway recently located by the Solstice Project (Marshall and Sofaer, 1988) that descends from the Kutz Canyon escarpment to the canyon floor (see Figures 29.5 and 29.6).

From Pueblo Alto to Kutz Canyon, the road lies within one corridor, with no evidence of bifurcations. For much of its length, it exists as two, and occasionally four, closely spaced, parallel roads.

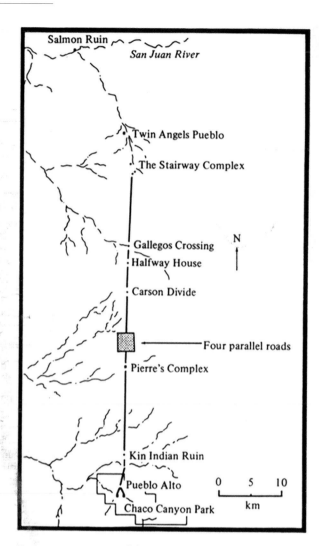

FIGURE 29.3. *Map of the Great North Road. (Copyright © 1986 Solstice Project.)*

The road's length and the complexity of its construction have led scholars to term it the "Great North Road."

The road traverses rolling, sagebrush country, where the only prominent natural features in view, and then only from rises, are the distant snowcapped mountains to the north. The only major topographic relief are the canyons at each end. With the possible exception of Pierre's Complex, there are no communities on the road's course from Pueblo Alto to Kutz Canyon. Two large complexes—Aztec and Salmon Ruin—lie to the northwest, 20 and 30 km beyond the road's terminus. Most of the outlying Chacoan communities are to the south, west and east of Chaco Canyon.

The road has been traced in numerous segments by aerial photography. On the ground, it has been intensively investigated from Chaco Canyon to Pierre's Complex and partially studied from there to Kutz Canyon. Its straight course and distinctive parallel segments have aided scholars in identifying and following it. Associated ceramic scatters and a number of unusual structural sites along its route have also aided its ground detection. Because earth and vegetation have refilled the road, only limited vestiges of it are visible on the ground today and sometimes only under particular lighting conditions.

Construction of the road involved primarily the removal of earth and vegetation. Extensive road cuts were made where the road crosses land elevations. Near the large community buildings of the canyon, several stairways were sculpted and large ramps were constructed. Several of the multiple roadways connecting these stairways and ramps with Pueblo Alto were curbed with masonry. Along one of these segments there is a curious linear groove cut into the bedrock. The stairway at Kutz Canyon, now largely collapsed, was built as a series of platforms which were supported by juniper posts and crossbeams and packed with earth (Marshall and Sofaer, 1988). The effort required for the road's construction testifies to the serious purpose that attended the decision to plan and execute it.

Considered from a utilitarian perspective, however, the road appears to be *overbuilt* and *underused*. Important features of the road—its extraordinary width and the redundancy of its routes—have no satisfactory functional explanation. The road averages 9 meters in width—wider than a modern two-lane road and far wider than any of the other prehistoric roads or trails of the Southwest outside of the Chaco cultural region. The width is greater than required for draft animals or wheeled vehicles. Since this culture had neither, the width seems especially excessive in practical terms.

Redundancy occurs in the multiple stairways heading out of the canyon, the four routes that converge on Pueblo Alto, and most particularly where the Great North Road is expressed, for a good fraction of its length north of Pierre's Complex, as a set of two parallel roads. In addition, at one location, a set of four "almost perfectly parallel" roads extending for 1.5 km is evident in aerial photography (Nials, 1983, p. 6.29) (Figure 29.4). Recent re-evaluation of the aerial imagery for the Solstice Project has revealed further portions of the road in previous gaps to the north of Pierre's Complex (G. Obenauf, 1986, unpublished report to the Solstice Project on re-evaluation of Bureau of Land Management aerial photography). Many of these segments consist of two parallel roads. (The new portions lie on the straight line determined by the sections found earlier and thus further emphasize the overall linearity of the road.) There is no satisfactory functional explanation for these redundant features. Yet the effort devoted to achieving them indicates they are not casual expressions of the Chaco culture.

Viewed from a utilitarian perspective, we would expect the Great North Road to connect Chaco Canyon with other major population centers. An examination of the structures along the Great North Road and its destination, however, does not appear to support the earlier functional interpretation of its development and use. The road, after leaving the ceremonial complex of Chaco Canyon, traverses the least developed region of the Chaco cultural area. The structures along the road are small in comparison with other outlying Chaco structures, and minute in comparison with those in Chaco Canyon. All of the structures contain less than six rooms, and most of them contain less than three. Only Pierre's Complex suggests a possible community.

Earlier maps and reports of the Chaco cultural region have assumed that this road goes to Twin Angels Pueblo (Kincaid, 1983, Figure 4.1) and then extends at a NNW bearing, to one or both of the large San Juan River communities of Salmon Ruin (Powers *et al.* 1983; Cordell, 1984) and Aztec (Morenon, 1977). There is, however, no ground inventory or aerial investigation that provides evi-

133

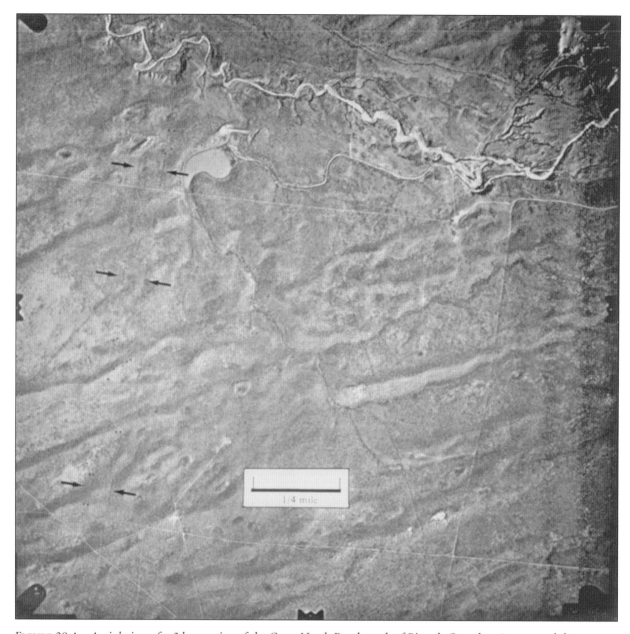

FIGURE 29.4. *Aerial view of a 2 km section of the Great North Road north of Pierre's Complex. Arrows and dots indicate the road's four parallel segments. (Other linear features are modern roads.) Bureau of Land Management aerial imagery, 1981.*

dence that, in fact, the road goes to these pueblos. Moreover, efficiency for travel and transportation of goods to Salmon Ruin and Aztec would dictate a more direct and easier route from Chaco Canyon—one further to the west. Instead, the road goes north and descends a nearly impassably steep slope of Kutz Canyon.

Twin Angels Pueblo is located in the Kutz Canyon badlands, 6 km from the road's apparent terminus (Carlson, 1966). It is a relatively small pueblo of 17 rooms—less than one-tenth the size of Salmon Ruin or Aztec. (We note with interest that, although there is no evidence of the continuation of the road to or near Twin Angels Pueblo, that site

134

lies only ½° east of north from the start of the road near Chaco Canyon. We cannot, at this point, rule out the possibility of a road relationship with this pueblo.)

A recent inventory of the Great North Road has produced no evidence that indicates extensive use for the transportation of economic goods (Stein, 1983). It is estimated that only 10% of the ceramics found on the Great North Road are from the San Juan River communities (Stein and Levine, 1983), giving scant evidence of significant trade with them. The absence of hearths and ground or chipped stone in the road inventory suggests there was little encampment along the road.

To summarize, the road's great width and parallel routes, its ephemeral practical use, and apparent terminus at an isolated badlands canyon fail to justify, in functional terms, the effort entailed in its construction. The road apparently goes "nowhere" and displays a level of effort far out of proportion to the meager tangible benefits that may have been realized from it. In many important respects, the road appears to be its own reason for development — an end in itself.

THE GREAT NORTH ROAD — PURPOSE

In the absence of a satisfactory functional explanation and practical destination for the Great North Road, and knowing what we do about Chaco and its interest in religious architecture, we posit that the primary purpose of the road may have been the expression of spiritual values. We will consider its direction to the north and its topographic direction with this in view. In addition, we will consider the sites along its course and their frequent location on distinctive land forms.

In the ceremonial architecture and astronomy of the Chaco culture the north-south axis is primary. Most of the great kivas have approximate north-south axes and the kivas generally have niches primarily located to the north (Reyman, 1976). The axes of two major ceremonial structures of Chaco Canyon, Pueblo Bonito and the great kiva, Casa Rinconada, are within ¼° of north (Williamson *et al.* 1977; A. Sofaer and R. M. Sinclair,

1984, unpublished survey). A bearing within ½° of north-south has been noted between two high ceremonial structures which are intervisible — Pueblo Alto and Tsin Kletzin (Fritz, 1978) — the former of which is itself aligned to the cardinals (Sofaer and Sinclair, 1986b). It is interesting to observe that Pueblo Alto and Pueblo Bonito are origin points in the canyon for the Great North Road. Just prior to the time of the road's construction, Alto was constructed; close to the time of the road's actual development, Bonito was greatly expanded and given its cardinal alignments (Lekson, 1984).

Seven noon-seasonal markings using shadow and light patterns on petroglyphs on Fajada Butte also involve the north-south axis. They occur within a few minutes of meridian passage of the sun, when the sun is due south, and thus involved a comparable interest in and knowledge of the north-south axis (Sofaer and Sinclair, 1986a).

The effort made by the Chacoans to construct the Great North Road with a bearing within ½° to 2° of true north is similar to the effort they made to involve the north-south axis in their large ceremonial constructions and the noon-seasonal markings on Fajada Butte. It is important to note that the road appears to deviate intentionally from astronomic north after Pierre's Complex in order to arrive at the dramatic edge of Kutz Canyon. Clearly the people of Chaco had the capability of directing the road to within ½° of north, and as noted above they did so in a 16 km segment. For the next 31 km of the road they departed from this bearing and struck, with a rigorously straight course, their direction to a large mound on the edge of Kutz Canyon. The purpose of this deviation appears to have been a blending of astronomic north and symbolic use of topographic features in a cosmographic expression.

The road's 2° angle change directs it straight from the cone-shaped mound at Pierre's Complex, El Faro, to the large Upper Twin Angels mound. This mound (Figure 29.5) is located on the edge of the steepest slope of Kutz Canyon, where the stairway descends to the canyon floor. The mound stands out prominently above the deeply eroded slope of the canyon wall (Figure 29.6); from 10 km

FIGURE 29.5. *Upper Twin Angels mound, looking north, seen from the edge of the canyon. A small shine-like structure is located on top. The road terminus and stairway are just out of the picture to the left. (Photograph by Anna Sofaer, Copyright © 1986 Solstice Project.)*

to the north, it is the only relief that extends above the southern horizon. These symmetrically shaped pinnacles, El Faro and Upper Twin Angels mound, while not very high, are the most distinctive prominences in the vicinity of the road corridor as it crosses the rolling northern terrain.

The straightness of the Great North Road has suggested that it was "laid out as a single unit" (Morenon, 1977), and the "chronological homogeneity" in the material culture associated with it has suggested "that it can be viewed as a single construction event" (Kincaid, Stein and Levine, 1983, p. 9.76). The sites adjacent to the road were built at the time of its construction, apparently in association with its construction and use.

Five isolated structures along the road are small low-walled units located on distinctive land forms such as pinnacles or ridge crests (Kincaid, 1983; Marshall and Sofaer, 1988). They resemble shrines of the historic Pueblo culture, which are similarly small, often in remote locations, and frequently on elevated land forms. Such a site was constructed on the top of the Upper Twin Angels mound.

A Pierre's Complex, almost all of the 27 structures are located on pinnacles, mesa tops, and steep ridge slopes (Powers *et al.* 1983). While it is the largest development on the road and three of its structures are similar in scale to some small-to-medium outlying Chacoan pueblos, it is atypical. About a third of the structures are isolated rooms or non-habitation sites. A recent report describes it as "a constellation of special-function architecture," the location of which "was probably

136

predetermined during the engineering of the North Road" (Stein, 1983, p. 8.9). This report further states: "indeed, arrangement of the major structures within the complex acted to preserve the bearing of the road and to 'receive' it into the community." These structures include a hearth construction on top of El Faro, from which there is extended visibility north and south.

Certain aspects of the ceramics associated with the Great North Road suggest the possibility of ceremonial activity on the road. There are several curious concentrations of shards along the road at locations isolated from structures and without evidence of nearby encampment. Unusually dense elongate ceramic scatters occur along the road several kilometers south of Pierre's Complex (Kincaid *et al.* 1983, p. 9.74). Along the isolated Kutz Canyon stairway, there is a concentration of ceramics (Marshall and Sofaer, 1988). The extensive quantity of broken ceramics at Pueblo Alto has suggested to some analysts the possibility of large ritual gatherings involving dispersal of food items and deliberate breakage of vessels (Judge, 1984; Toll, 1984). The ceramics along the Great North Road (Kincaid *et al.* 1983) and at Pueblo Alto (Toll, 1984) have a significantly higher proportion of jars and non-utility ware than the ceramics at a typical Chacoan site. The road's enigmatic ceramic concentrations, the possibility of ceramic-related rituals at Pueblo Alto, and the character of the road ceramics suggest the possibility of ceremonial activity associated with ceramics on the road.

OTHER CHACO ROADS

Many of the other Chaco roads also exhibit non-utilitarian features and suggest cosmological purposes. For instance, the other roads are as wide and straight, and show no evidence of frequent use. Long linear grooves were cut into the bedrock along certain roads. The principal road to the south, the South Road, has a segment of parallel roads. The ceramics on the other roads share the same non-utility ratio as the Great North Road. Small isolated structures resembling historic shrines have been found so frequently along the roads that they

are now used as a means of predicting the presence of a road (Kincaid *et al.* 1983, p 9.16).

Recent road inventories have discounted several earlier postulated road connections between Chaco Canyon and major communities (Nials, Stein and Roney, 1988). Only a few such roads have been verified. Where certain roads do articulate with large communities, they appear to be only interconnecting avenues between nearby outlying communities or to link structures within the community, such as the large public building and the great kiva. Sometimes they appear to represent only a formalized entranceway to ceremonial locations in the community. Where the road articulates with public buildings, there is frequently evidence of large ceremonial earthen architecture: ramps, circular mounds, and platform mounds (J. R. Stein, 1983, private communication). The roads in this vicinity are usually wider and often curbed with masonry (see Figure 29.2). The pecked grooves and evidence of fire on ramps, burnt structures, elevated fire boxes, and fire pits warrant further investigation for possible ceremonial significance. At Pueblo Alto, very large fire pits are located at the road's entrance points.

Some roads lead *only* to topographic features such as pinnacles, springs, or lakes. One major road, the Ashlislepah Road, which runs from Peñasco Blanco in Chaco Canyon 12 km to the northwest, connects with no other communities. It articulates, instead, with a group of cisterns, where there is a small, apparently non-utilitarian site, and then appears to terminate at now-dry Black Lake (Marshall and Sofaer, 1988). Another road, which runs from the community of Kin Ya'a and the southern terminus of the South Road 7 km to the south appears to terminate near the base of massive Hosta Butte, one of the highest and most prominent natural formations in the San Juan Basin (Nials *et al.* 1988).

ETHNOGRAPHIC BACKGROUND

Historic Pueblo cosmology and ceremony may afford insights into the religious considerations underlying construction of the Great North Road and other Chaco roads. Here we find frequent sym-

FIGURE 29.6. *North slope of Kutz Canyon seen from the canyon floor. The arrow indicates the top of the stairway. (Photograph by Anna Sofaer, Copyright © 1986 Solstice Project.)*

bolic use of straight roads, mythic and ceremonial journeys to and from the north and the "middle place," and attention to prominent topographic features as elements of a spiritual landscape. There is even evidence of emblematic use of parallel roads and pecked grooves.

There are many symbolic uses of roads in Pueblo ritual and myth. "Road" translates as "channel for the life's breath" in Tewa, a Pueblo language (A. Ortiz, 1987, private communication). "Life is a road; important spirits are...keepers of the roads, the life roads. All spirits or sacrosanct persons have a road of cornmeal or pollen sprinkled for them where their presence is requested" (Parsons, 1939, pp. 17-18). These roads can represent the road traveled by the people to the middle place from the shipapu, the place where they emerged from the worlds below (Parsons, 1939, pp.

310, 363). Sometimes the road is for the spirits of the dead to return to the shipapu (White, 1942, p. 177).

For the Keresan Pueblos, north is where Iyatiku, the mother of all, resides at the shipapu. An account of Keresan cosmology describes the importance of north and the road to the north (White, 1960). When the people came out from the worlds below "they stayed near the opening at Shipapu for a time, but it was too sacred a place for permanent residence, so Iyatiku told them they were to migrate to the south." They moved south and stopped at a place where they lived for a long time.

When people died, their bodies were buried, but their souls went back to Shipapu, the place of emergence and returned to their mother in the fourfold womb of the earth... So every year, now, the souls of the dead come back to the pueblos of the living and visit their relatives and

138

eat the food that has been placed for them on their graves and on the road to the north.

This "road to the Shipapu" is described in another report as "crowded with spirits returning to the lower world, and spirits of unborn infants coming from the lower world" (Stevenson, 1894, p. 67). This and other roads are frequently described as "straight" (Stevenson, 1894, pp. 31, 41, 145).

When a person dies in the Keresan and Tanoan pueblos, the officiant takes offerings that represent that person's soul to the north and deposits them in a canyon or a mesa crevice (White, 1973, p. 137). Ceramic vessels are frequently broken in rituals related to the dead (Parsons, 1939, pp. 72, 77; Ortiz, 1969, p. 54). Sometimes a vessel containing food which is "the last meal of the deceased" is put on the road to the north or sometimes it is "killed" (broken at the rim) and then thrown by the officiant "out to the north, the direction in which the soul…travels toward Shipapu" (White, 1942, p. 177).

Traditionally, the Pueblo people re-enact the creation and emergence events, especially at important solar times. As part of these ceremonies, they make ritual journeys to certain mountains, canyons, caves and lakes—places they regard as shipapu openings (Ellis and Hammack, 1968, pp. 31, 33; Ladd, 1983). These journeys may be as long as 500 km to and from the pueblo. Along the route, the ceremonialists leave offerings at shrines which are located on distinctive land forms, such as buttes, cone-shaped hills, ravines and springs. From a Keresan pueblo on the south edge of the prehistoric Chaco region, ceremonialists packed their burros with solar offerings and traveled north, stopping first at Chaco Canyon (Ellis and Hammack, 1968, p. 32). They made offerings at a shrine on the south side of the canyon and then travelled to a shrine at Jackson Butte and finally to the shipapu, a small lake or spring in the San Juan Mountains. In Pueblo culture, the mountains are where the cloud beings, the spirits of the dead, reside (Parsons, 1939, pp. 172, 173).

For Jemez, a Tanoan pueblo, north is "spiritually indicative of the mythical and ancestral homeland" (Weslowski, 1981, p. 123), and the place of emergence is in the mountain range to the north of the pueblo. One of their most sacred shrines is located on its prominent peak. There the "underworld chiefs make a pilgrimage…every June to begin the summer series of rain retreats and ceremonies" (Weslowski, 1981, p. 117). In the emergence and migration of this pueblo, the leader Fortease, upon emergence from the shipapu, chooses the direction "towards the south" and then makes four roads for the people to travel on in search of their place of settlement (Parsons, 1925, pp. 137, 138). Fortease is reported to have made the roads by "clearing away the brush." Reference to two parallel roads occurs in Tanoan cosmology (Ortiz, 1969, p. 57):

> *True to the underlying message of the origin myth…the Tewa do begin and end life as one people. The term they use for the life cycle is poeh, or "path," after the two different migration paths the moieties followed after emergence. Thus, at the beginning of life there is a single path for all Tewa…it divides into two parallel paths and continues in that way until the end of life. At death the paths rejoin again and become one, just as the moieties rejoined in the myth of origin.*

At the Zuñi Pueblo, a pilgrimage is conducted every four years at summer solstice by 50 religious leaders to a lake, the Zuñi "village of the gods," the place where the spirits of all Zuñis go after death (E. R. Hart, 1985, unpublished manuscript: "The Zuñi Indian tribe and title to Kolhu/wala:wa (Kachina Village)). Fires are lit along the route by one of the participants, the Zuñi Fire God. Another important pilgrimage, to the Zuñi Salt Lake, is on roads that have been described as very straight and with shrine-like sites similar to those on the Chaco roads (Kelley, 1984). Although for the Zuñi, these sacred lakes and the origin place are not located to the north, north is associated with the "undermost" of the below worlds (Stevenson, 1904, p. 25) and has primacy in the ordering of ceremonial events and religious leadership (Cushing, 1979, pp. 188-90). In the prayers and chants telling of their emergence and migration to the middle (i.e. Zuñi), reference is made to four parallel roads: "Hither towards Itiwana (the middle) I saw four roads going side by side." (Bunzel, 1932a, p. 717). One Zuñi ceremony includes breakage by the religious leaders of ceramic vessels throughout the pueblo (Cushing, 1979, p. 321).

Long linear grooves are cut into the mesas near two present-day pueblos. Contact with these grooves is reported to help diagnose sickness in curing ceremonies, to help people to regain their strength and to help persons "to cease yearning for the dead or absent and to keep them from returning in their dreams" (Parsons, 1939, p. 449). There are many references to running in Pueblo ceremony and myth, sometimes on north-south roads (Parsons, 1925, p. 119) and sometimes symbolic of emergent events on north-south parallel courses (Dutton and Marmon, 1936, p. 12); and in certain instances on ritually swept east-west tracks to aid the journey of the sun (Parsons, 1939, p. 547).

In several reports, canyons or deep holes are seen in myths or in the actual topography as leading to the shipapu (E. R. Hart, unpublished manuscript). Symbolic ladders connect with the underworld (Bunzel, 1932b, pp. 589, 590; Cushing, 1979, p. 132). At one Keresan Pueblo, the shipapu is described as (Lange, 1959, p. 416) "the Lagune…to the north, beyond the 'Conejos'…very round and deep. Many streams flow into it, but it has no issue. Out of this Lagune came forth the Indians and in it dwells 'Te-tsha-aa', our mother… The good ones return to her."

The "middle place," so important in Pueblo cosmology, is seen as the place of the convergence of the cardinal directions and the nadir and zenith directions (Ortiz, 1972, p. 142). Where these directions join is the sacred center for Pueblo people. This place is sometimes symbolically conveyed in the joining of ritual roads at the pueblo center (Goldfrank, 1962, p. 47). It is interesting to note that the cardinal directions in the Chaco architecture, and the north-south axis of the Great North Road, merge at a central ceremonial complex of the vast Chaco cultural province, Pueblo Bonito.

CONCLUSION

The Great North Road embodies many non-utilitarian aspects and has no clear practical destination. It displays a level of effort in its engineering and construction that is far out of proportion to any material benefits that could be realized from it. Its direction to the north and its linkage to the middle place of Chaco Canyon find echoes in much of the tradition of the historic Pueblos, where roads, and especially a road to the north, have intense symbolic value. We conclude that the Great North Road was conceived, in harmony with much of Chaco architecture, as a cosmographic expression uniting the Chaco world and its works with its spiritual landscape.

ACKNOWLEDGMENTS

The Solstice Project investigations have relied heavily on the comprehensive Bureau of Land Management "Chaco Roads Project Phase I" (Kincaid, 1983), which summarizes earlier pioneering studies of the Chaco roads, including the initial observations of the Great North Road by Pierre Morenon. We are grateful to Margaret Obenauf for her re-evaluation of aerial photography that led to further evidence for the Great North Road. We want to thank John Stein for sharing his early ideas regarding the possibility of cosmographic expression in the Chaco roads and his thorough knowledge of them. Conversations with Alfonso Ortiz, Fred Eggan, John Roney, Fred Nials and Chris Kincaid have been particularly helpful to this study. Our profound thanks go to a friend who edited this paper with remarkable insight and generosity, and who, most remarkably, insists on remaining anonymous.

REFERENCES

Akins, N. J. and J. D. Schelberg
   1984. "Evidence for organizational complexity as seen from the mortuary practices at Chaco Canyon," *in Recent Research on Chaco Prehistory*, eds. W. J. Judge and J. D. Schelberg, pp. 89-102. U.S. Department of the Interior, National Park Service, Albuquerque.

Bunzel, R. L.
1932a. *Zuñi Ritual Poetry*. 47th Annual Report. Bureau of American Ethnology, pp. 611-835. Smithsonian Institution, Washington, D.C.

Bunzel, R. L.
1932b. *Zuñi Origin Myths*. 47th Annual Report, Bureau of American Ethnology, pp. 545-609. Smithsonian Institution, Washington, D.C.

Carlson, R. L.
1966. "Twin Angels Pueblo." *American Antiquity* 31, (5), 676-82.

Cordell, L. S.
1984. *Prehistory of the Southwest*. Academic Press, Orlando, Fla.

Cushing, F. H.
1979. *Zuñi, Selected Writings of Frank Hamilton Cushing*, ed. J. Green. University of Nebraska Press, Lincoln.

Dutton, B. P. and M. A. Marmon
1936. "The Laguna Calendar." *University of New Mexico Bulletin, Anthropological Series* 1, (2), 3-21.

Ellis, F. H. and L. Hammack
1968. "The Inner Sanctum of the Feather Cave," in *American Antiquity* 33, (1), 25-44. Academic Press, Orlando, Florida.

Fritz, J. M.
1978. "Paleopsychology Today," in *Social Archaeology: Beyond Subsistence and Dating*. Academic Press, Orlando, Florida.

Goldfrank, E. C. (ed.)
1962. *Isleta Paintings*. Bureau of American Ethnology Bulletin 181. Smithsonian Institution, Washington, D.C.

Judge, W. J.
1984. "New Light on Chaco Canyon," in *New Light on Chaco Canyon*, ed. D. G. Noble, pp. 1-12. School of American Research Press, Santa Fe.

Kelley, K. B.
1984. *Historic Cultural Resources in the San Augustine Coal Area*. Draft on file at Bureau of Land Management, Socorro District, New Mexico, June 1984.

Kincaid, C. (ed.)
1983. *Chaco Roads Project Phase I, A Reappraisal of Prehistoric Roads in the San Juan Basin*. U.S. Department of the Interior, Bureau of Land Management, Albuquerque.

Kincaid, C., J. R. Stein, and D. F. Levine
1983. "Road Verification Summary," in *Chaco Roads Project Phase I, A Reappraisal of Prehistoric Roads in the San Juan Basin*, ed. C. Kincaid, pp. 9.1-9.78. U.S. Department of the Interior, Bureau of Land Management, Albuquerque.

Ladd, E.
1983. "Pueblo Use of High-Altitude Areas: emphasis on the A: Shiwi," in *High Altitude Adaptations in the Southwest*, ed. J. C. Winters, pp. 168-76. U.S. Forest Service, Southwest Region Report no. 2, Albuquerque.

Lange, C. H.
1959. *Cochiti: A New Mexico Pueblo, Past and Present*. University of Texas Press, Austin, reprinted in 1968 by Southern Illinois University Press, Carbondale.

Lekson, S. H.
1984. *Great Pueblo Architecture of Chaco Canyon, New Mexico*. U.S. Department of the Interior, National Park Service, Albuquerque.

Marshall, M. P. and A. Sofaer
1988. *Report on the Solstice Project Archaeological Investigations in the Chacoan Province, New Mexico*. (To be published.)

Marshall, M. P., J. R. Stein, R. W. Loose and J. E. Novotny
1979. *Anasazi Communities of the San Juan Basin*. Public Service Company of New Mexico, Albuquerque.

Morenon, E. P.
1977. *A View of the Chacoan Phenomenon from the "Backwoods": A Speculative Essay*. On file, Archaeological program of the Institute of Applied Sciences, North Texas State University, Denton, Texas.

Nials, F. L.
1983. "Physical Characteristics of Chacoan Roads," in *Chaco Roads Project Phase I, A Reappraisal of Prehistoric Roads in the San Juan Basin*, ed. C. Kincaid, pp. 6.1-6.51. U.S. Department of the Interior, Bureau of Land Management, Albuquerque.

Nials, F. L., J. R. Stein, and J. R. Roney
1988. *Chacoan Roads in the Southern Periphery: Results of Phase II of the Bureau of Land Management Chaco Roads Study*. Department of the Interior, Bureau of Land Management, Albuquerque (in press).

Ortiz, A.
1969. *The Tewa World: Space, Time, Being, and Becoming in a Pueblo Society.* University of Chicago Press.

Ortiz, A. (ed.)
1972. *New Perspectives on the Pueblos.* University of New Mexico Press, Albuquerque.

Parsons, E. C.
1925. *The Pueblo of Jemez.* Papers of the Phillips Academy Southwestern Expedition, 3. Yale University Press, New Haven.

Parsons, E. C.
1939. *Pueblo Indian Religion.* University of Chicago Press.

Powers, R. P., W. B. Gillespie, and S. H. Lekson
1983. *The Outlier Survey, A Regional View of Settlement in the San Juan Basin.* U.S. Department of Interior, National Park Service, Albuquerque.

Reyman, J. E.
1976. "The Emics and Etics of Kiva Wall Niche Location." *Journal of the Steward Anthropological Society* 7, (1): 107-29.

Sofaer, A., V. Zinser, and R. M. Sinclair
1979. "A Unique Solar Marking Construct." *Science* 206, 283-91.

Sofaer, A., R. M Sinclair, and L. E. Doggett
1982. "Lunar Markings on Fajada Butte, Chaco Canyon, New Mexico," in *Archaeoastronomy in the New World,* ed. A. F. Aveni, pp. 169-86. Cambridge University Press.

Sofaer, A. and R. M. Sinclair
1986a. "Astronomical Markings at Three Sites on Fajada Butte," in *Astronomy and Ceremony in the Prehistoric Southwest,* eds. J. Carlson and W. J. Judge. Maxwell Museum Technical Series, University of New Mexico, Albuquerque.

Sofaer, A. and R. M. Sinclair
1986b. "Astronomic and Related Patterning in the Architecture of the Prehistoric Chaco Culture of New Mexico," *Bulletin of the American Astronomical Society* 18, (4), 1044-5.

Stein, J. R.
1983. "Road Corridor Descriptions," in *Chaco Roads Project Phase I, A Reappraisal of Prehistoric Roads in the San Juan Basin,* ed. C. Kincaid, pp. 8.1-8.15. U.S. Department of the Interior, Bureau of Land Management, Albuquerque.

Stein, J. R. and D. F. Levine
1983. "Documentation of Selected Sites Recorded During the Chaco Roads Project," in *Chaco Roads Project Phase I, A Reappraisal of Prehistoric Roads in the San Juan Basin,* ed. C. Kincaid, pp. C.1-C.64. U.S. Department of the Interior, Bureau of Land Management, Albuquerque.

Stevenson, M. C.
1894. *The Sia.* 11th Annual Report of the Bureau of American Ethnology, pp. 3-157. Smithsonian Institution, Washington, D.C.

Stevenson, M. C.
1904. *The Zuñi Indians.* 23rd Annual Report, Bureau of American Ethnology, pp. 3-634. Smithsonian Institution, Washington, D.C.

Toll, H. W.
1984. "Trends in Ceramic Import and Distribution in Chaco Canyon," in *Recent Research on Chaco Prehistory,* eds. W. J. Judge and J. D. Schelberg, pp. 115-36. U.S. Department of the Interior, National Park Service, Albuquerque.

Vivian, R. G.
1983. "Identifying and Interpreting Chacoan Roads: an Historical Perspective," in *Chaco Roads Project Phase I, A Reappraisal of Prehistoric Roads in the San Juan Basin,* ed. C. Kincaid, pp. 3.1-3.20. U.S. Department of the Interior, Bureau of Land Management, Albuquerque.

Weslowski, L. V.
1981. "Native American Land Use Along Redondo Creek," in *High Altitude Adaptation Along Redondo Creek,* eds. C. Baker and J. C. Winter. Office of Contract Archaeology, University of New Mexico, Albuquerque.

White, L. A.
1942. *The Pueblo of Santa Ana, New Mexico. Memoirs of the American Anthropological Association* 60, Menasha, Wisconsin.

White, L. A.
1960. "The World of the Keresan Pueblo Indians," in *Culture in History: Essays in Honor of Paul Radin,* ed. S. Diamond, pp. 53-64. Columbia University Press, New York.

White, L. A.
1973. *The Acoma Indians.* Rio Grande Press, Glorieta, New Mexico.

Williamson, R. A., H. J. Fisher, and D. O'Flynn
1977. "Anasazi Solar Observatories," in *Native American Astronomy,* ed. A. F. Aveni, pp. 203-18. University of Texas Press, Austin.

Computer
Restoration
of the
Sun Dagger
Site

8

Changes in Solstice Marking at the Three-Slab Site,

New Mexico, U.S.A.

"Changes in Solstice Marking at the Three-Slab Site" was first published in Archaeoastronomy *(Volume 21, No. 15) in 1990. At that time, Anna Sofaer and the Solstice Project were located in Washington, D.C. Rolf M. Sinclair was with the National Science Foundation in Washington, D.C.*

Anna Sofaer
and
Rolf M. Sinclair

WE HAVE RECENTLY OBSERVED A MAJOR CHANGE in the summer solstice light marking at a prehistoric Pueblo site on Fajada Butte in Chaco Canyon, New Mexico, where three large stone slabs collimate sunlight onto two spirals carved on a cliff face. The light patterns so formed have marked the summer and winter solstices and the equinoxes (Sofaer *et al.* 1979), as well as the lunar standstill cycle (Sofaer *et al.* 1982). The center slab casts the shadows that shape the top and left side of the summer solstice pattern (Sofaer *et al.* 1979, Figure 12). This slab also forms part of the light patterns at winter solstice and equinox.

In recordings made since its discovery in 1977, the summer solstice pattern would commence before midday with the appearance of a spot of light at the top edge of the larger spiral. The light pattern would then lengthen into a sharply pointed form 2 to 3 centimeters wide and descend through the center of the spiral. The total duration of this marking was 18 minutes (Figure 1; see also Sofaer *et al.* 1979, Figures 7a-e).

The center slab has recently shifted approximately 6 cm perpendicular to its face. As a result of this shift, the light pattern at summer solstice now appears first as a small streak of light about 15 cm above the spiral. The light form then lengthens into an irregular shape 7 cm wide (Figure 1). The total duration of this light form is more than 40 minutes. The distinctive visual quality of the previous sharply formed and slender pattern bisecting the spiral is lost. Preliminary results of our analysis of the photographic record of the light pattern show that most or all of the change occurred since 1987.

We had earlier noted that disturbances at the site had occurred in the years since its discovery (Sofaer 1982; *Congressional Record 1982*). Prelimi-

FIGURE 1. *The irregular shape of the light pattern at summer solstice, compared with the previous sharply formed and slender pattern. (Photograph by Anna Sofaer, Copyright © Solstice Project.)*

nary examination of the photographic record of the rock slabs since 1978 reveals a lowering in the level of earth at the base of the slabs of up to 25 cm between 1978 and 1989. We speculate that the recent shift in the center slab is connected with this loss of supporting and stabilizing material.

Concerned over the fragility and vulnerability of the site (Sofaer *et al.* 1979, 291), we have docu-mented its physical attributes and astronomical markings with extensive photographic and pho-togrammetric measurements (Curry *et al.* 1987). We are now preparing in cooperation with Rensse-laer Polytechnic Institute an interactive computer graphics model which will serve as an archival duplication of the site as well as a research tool (Bordner 1989; Sofaer *et al.* 1989).

BIBLIOGRAPHY

Bordner, Karen
1989. "Computer Graphics Unlock Mysteries of the Past," *Rensselaer,* March issue, 4-7; *Congressional Record.* 1982 Issue of 9 December, p. S14283.

Curry, S., D. Fair, D. Encinias, A. Sofaer and R. M. Sinclair
1987. "Mapping the Sun Dagger," American Congress on Surveying and Mapping — American Society for Photogrammetry and Remote Sensing Annual Convention: *Technical Papers,* ii, 1-7.

Sofaer, A.
1982. "Protective Measures for the Sun Dagger Site, Chaco Canyon National Historic Park, fiscal year 1983." Unpublished memorandum to Senator Pete V. Domenici.

Sofaer, A., R. M. Sinclair and E. Brechner
1989. "Computer Graphics Model of the Three-Slab Site on Fajada Butte, New Mexico," *Bulletin of the American Astronomical Society,* xxi, 1210.

Sofaer, A., R. M. Sinclair and L. Doggett
1982. "Lunar Markings on Fajada Butte, Chaco Canyon, New Mexico," in *Archaeoastronomy in the New World,* ed. by A. F. Aveni, 169-81, Cambridge, U.K.

Sofaer, A., V. Zinser and R. M. Sinclair
1979. "A Unique Solar Marking Construct," *Science,* ccvi, 283-91.

9

The Sun Dagger Interactive Computer Graphics Model:

A Digital Restoration of a Chacoan Calendrical Site

This paper was first presented to the Tenth Biennial Southwest Symposium in Las Cruces, New Mexico, in January 2006. Alan Price is associated with Ohio State University; James Holmlund and Joseph Nicoli are with Western Mapping Company, and Andrew Piscitello is with Aero-Metric Company.

Anna Sofaer
Alan Price
James Holmlund
Joseph Nicoli
and
Andrew Piscitello

A N INTERDISCIPLINARY TEAM IN 2006 produced an interactive computer graphics model that precisely replicates the astronomical functioning of an ancient calendrical site, the Sun Dagger, of Chaco Canyon, New Mexico. This digital restoration in an interactive format allows further exploration research of the structure's geometry and the process of its development.

At the Sun Dagger site, which Anna Sofaer rediscovered in 1977, three upright sandstone slabs cast precise light and shadow patterns on two spiral petroglyphs, recording the summer and winter solstices, the equinoxes, and the 18.6 year lunar cycle (Sofaer *et al.* 1979; Sofaer *et al.* 1982; Sofaer and Sinclair 1987; Sinclair *et al.* 1987, Figures 1–4). Certain of today's Pueblos, descendents of the Chacoan culture, regard the Sun Dagger as a sacred site, expressing the spiritual significance of the sun and the moon.[1]

Over the ten years following Sofaer's 1977 finding, the Solstice Project collected timed photographic records of the light and shadow formations of the Sun Dagger site.[2] In 1989, the project discovered that disturbances at the site had destroyed its solar marking functions. Owing in large part to the site's attractiveness to tourists and investigators, the middle of the three slabs had pivoted two to five centimeters from its originally recorded position (Palca 1989; Sofaer and Sinclair 1990; Trott *et al.* 1989).

In 2005, combining state-of-the-art Global Positioning System (GPS) and 3D Light Detection and Ranging (LiDAR) techniques with photogrammetric measurements collected in 1984, a team that included an archaeoastronomer, a geodesist, a computer graphics specialist, and several mapping specialists produced two high-resolution 3D computer models of the 1984 and the 2005 positions of the slabs (Nicoli *et al.* 2006).

Submitted for publication in Acts of History: Ritual, Landscape, and Historical Archaeology, *William Walker and Maria Nieves Zedeño, editors (University Press of Colorado, Boulder, 2008).*

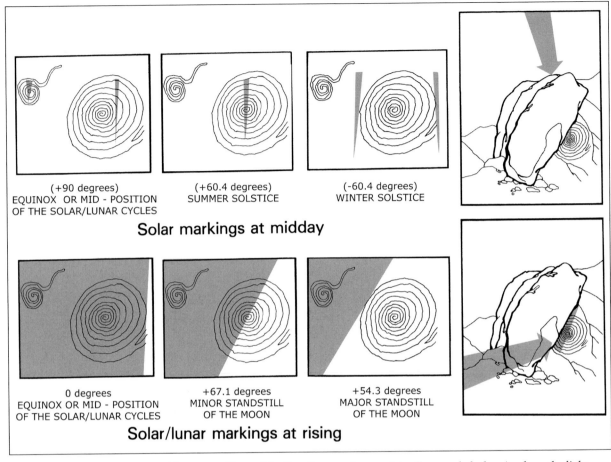

(+90 degrees)
EQUINOX OR MID - POSITION
OF THE SOLAR/LUNAR CYCLES

(+60.4 degrees)
SUMMER SOLSTICE

(-60.4 degrees)
WINTER SOLSTICE

Solar markings at midday

0 degrees
EQUINOX OR MID - POSITION
OF THE SOLAR/LUNAR CYCLES

+67.1 degrees
MINOR STANDSTILL
OF THE MOON

+54.3 degrees
MAJOR STANDSTILL
OF THE MOON

Solar/lunar markings at rising

FIGURE 1. *Diagram of solar and lunar markings of the Sun Dagger site as originally recorded, showing how the light comes onto the spirals from the sun's passage above the slabs and from the sun's and moon's rising positions. (Copyright © Solstice Project.)*

Integrating these models produced a virtual restoration of the site that simulates the original interaction of sunlight and moonlight with the slabs and petroglyphs. This virtual restoration serves as an archival record of the site, an interactive research tool, and an educational resource.

THE SUN DAGGER SITE

Near the top of 135 meter high Fajada Butte at the south entrance of New Mexico's Chaco Canyon, three sandstone slabs lean against a southeast facing cliff (Figures 1, 4). Each measures two to three meters high and weighs two thirds to one and a third tons. The openings between the slabs are to the south-southeast. Of the two spiral petroglyphs pecked into the rock face behind the slabs, one, forty-one centimeters across, has nine and half turns; the second, to its left and thirteen centimeters across, has two and a half turns.

Solar Light Patterns: The slabs cast vertical light patterns onto the cliff each day in the hours of late morning to near solar noon. These patterns form markings on the two spirals that are distinctive to the solstices and equinoxes. At summer solstice the opening between the middle and the eastern slab forms a slim dagger of light. It begins forty-seven minutes before solar noon in the top turn of the large spiral as a small spot of light that lengthens to a dagger shape bisecting the spiral. It descends through and off the spiral eighteen minutes after its first appearance. Four days before or after the summer solstice, a slight rightward shift of the light dagger—of about 2 millimeters—occurs in its position in relation to the center of the large spiral. In four weeks it is 3.2 centimeters to the right of the center. Through the following weeks and months, the light dagger descends through the spiral in positions farther and farther to the right.

Soon after summer solstice a second light dagger, formed by the opening between the middle and the western slab, joins the first. It also descends in a vertical course along the cliff face each day and farther to the right as the seasons progress. The second light dagger bisects the small spiral at equinox one hour and ten minutes before solar noon.

At winter solstice the two vertical light daggers descend over the course of three hours. When the second dagger matches the first in length, one hour and forty-one minutes before solar noon, both are on the outer edges of the large spiral. At this time of the sun's lowest passage, the light daggers appear to frame the spiral—bisected by the sun during its highest passage, at summer solstice—now empty of light.

Lunar and equinox light patterns: The eastern slab creates shadow patterns that record the extremes of the moon in its 18.6 year cycle—the northern major and minor lunar standstill positions—and the equinox, or the mid position of the sun or moon. The inner (or northern) edge of the eastern slab casts a shadow on or adjacent to the large spiral as the sun rises or the moon rises (during phases of its night time rise) at azimuths between approximately 54 and 90 degrees. In the year of the major lunar standstill, when the moon rises at its most northern position in its 18.6 year cycle at an azimuth close to 54.3 degrees, the eastern slab casts a shadow tangent to the left edge of the large spiral. Nine and a third years later, when the moon is at its northern minor standstill position at an azimuth close to 67.1 degrees, a shadow from this same slab diagonally bisects the large spiral. Twice yearly at the equinox, when the sun rises—or when the moon at the *midposition* of its cycle rises—directly east (close to 90 degrees), the same slab casts a shadow onto the right edge of the large spiral.[3]

It appears that the Chacoans gave distinctive emphasis to the lunar patterns by their alignment of two pecked grooves with the shadows of the minor and major standstill moons (Sofaer and Sinclair 1987 Figure 4.3, and in this paper Figure 3b). It is also of interest that the ten turns of the left side of the spiral that the shadows cross, year by year, during the progression of the moon from its northern minor to its northern major standstill correspond closely to the nine and a third years of this journey by the moon. In addition, the nineteen turns of the spiral from its left to right edges may symbolize the full lunar standstill cycle of 18.6 years.[4]

In sum, the three large sandstone slabs at the Sun Dagger site mark the extremes and midpositions of both the solar and lunar cycles on the same large spiral and equinox on the smaller spiral as well. The site reveals significant coordination in the markings. The Chacoans created solar and lunar markings that repeatedly fall on the center and outer edges of the large spiral. These findings raise a number of intriguing research questions. How critical are the relationships of the slabs' positions and shapes to these light and shadow formations? What activities and process did the Chacoans conduct at the site? How might they have placed and shaped the spirals and the cliff face on which they are carved, and shaped the slabs and adjusted their positions to achieve their results?[5]

DEVELOPING AN ARCHIVAL RECORD

Beginning in 1978, the Solstice Project set out to develop an archival record of the Sun Dagger site and envisioned a three dimensional computer graphics model for research. In fact, a quarter of a century would pass before the technology existed to accurately replicate the site in such a model. Yet early on, the project recognized that, beyond providing an archival record of the fragile site, a 3D computer graphics model would allow interactive experiments to further test the sensitivity of the site's geometry to the positions of the sun and the moon. These experiments in turn would afford

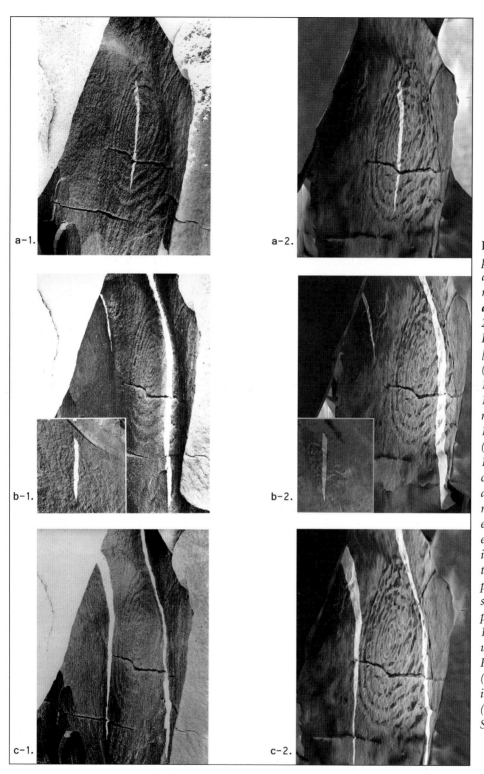

FIGURE 2: *Pairs of 1978 photographed images and 2006 registered model simulated images: **a**, summer solstice (June 26,1978, 11:13:15 am Local Apparent Time [LAT]); **b**, equinox (September 21, 1978, 10:50 am [LAT]; insert 10:52 am [LAT], not noted in Sofaer et al. 1979); **c**, winter solstice (December 22,1978, 10:22:15 am [LAT], corrected from Sofaer et al. 1979).(The correction noted here from Sofaer et al. 1979 is based on an error recently identified in the original reading of the winter solstice 1978 photographic contact sheets and log.) In each pair, the exact time of the 1978 recorded photo was used in the simulation. Photos by Karl Kernberger (1978) and simulated images by Alan Price (2006), (Copyright © Solstice Project).*

modern researchers the chance to gain insights into the process and concepts behind the Sun Dagger's original development.

Early Efforts.

Between 1978 and 1987, the Solstice Project collected comprehensive, timed photographic records of the light and shadow formations of the Sun Dagger site.

The project first contracted with photographer Karl Kernberger to record the light patterns on the spiral petroglyphs in an extensive series of precisely timed images. Near the twenty-first of each month of the 1978 solar cycle, he used a Hasselblad camera with a fifty millimeter lens to photograph from a single position, every thirty seconds, the changing light patterns through their duration.

In 1980, the Solstice Project photographed the lunar standstill markings on the spirals, using the sun to simulate the rising minor standstill moon's shadow and a laser to simulate the moon's shadow before it reached its major standstill. In 1987, when the moon had reached this position, the Project photographed the moon's shadow cast by the eastern slab (Sinclair et al. 1987, see in this paper Figure 3 c[1]). Between 1979 and 1987 the Project had also collected time lapse images with sixteen millimeter film of the light patterns on the spirals at the solstices and the equinox and recorded extensive measurements and documentation of other shadow and light formations, as well as the shadow casting edges of the rock slabs.

In addition to collecting these photographic records, and because of the extreme fragility of the site's soft sandstone slabs and spirals, the project contracted in 1981 with the engineering firm Koogle and Pouls to produce terrestrial photogrammetric measurements of the site as an archive and as data for a computer graphics model. The exactness of certain of the slab positions and shapes required to create the light patterns demanded fine precision in the construction of an archival computer model. For example, a one centimeter movement of the eastern slab on an approximate north-south axis would create a two centimeter displacement of the lunar light patterns on the large spiral; a one to two centimeter movement of the top surface of the eastern slab on an approximate east-west axis would create a similar displacement (or blockage) of the summer solstice light dagger. Because the light dagger itself is only two centimeters wide, such a change would destroy the astronomical accuracy of the marking.

Disturbances of the Site.

In 1981, the Solstice Project discovered, and then photographed, numerous disturbances and abuses to the Sun Dagger site: graffiti on the eastern slab, a beer can between the middle and eastern slabs, movement of several smaller rocks of the site, signs of removal of soil at the base of the middle slab, and the loss of material along the edge of the eastern slab that cast the lunar markings (Sofaer 1982).[6]

In response to this documentation, the U.S. Congress allocated $100,000 to the National Park Service (NPS) for the development of a computer graphics model of the site. Congress proclaimed the Sun Dagger site a national treasure and recommended its thorough protection by the NPS (U.S. Congress 1981). The NPS contracted with the consulting firm Ibarr, Inc., to develop a computer graphics model of the site. Ibarr contracted with two engineering firms—Aero-Metric and Dennett, Muessig, and Ryan—to produce a second set of terrestrial photogrammetric measurements of the site in 1984 but it did not develop a model of the site from the resulting data.

In 1989 the Solstice Project discovered significant changes at the Sun Dagger site caused by natural erosion, greatly accelerated by tourist traffic and investigative activity. The middle slab, critical to the shaping of all the solar light patterns, had pivoted about two to five centimeters from its documented position in 1978. Photo documenting of the effect of this disturbance by the project and the NPS showed substantial changes: The slim, pointed summer solstice dagger of 1978 was now a wide band of light; the two winter solstice light daggers did not bracket the large

153

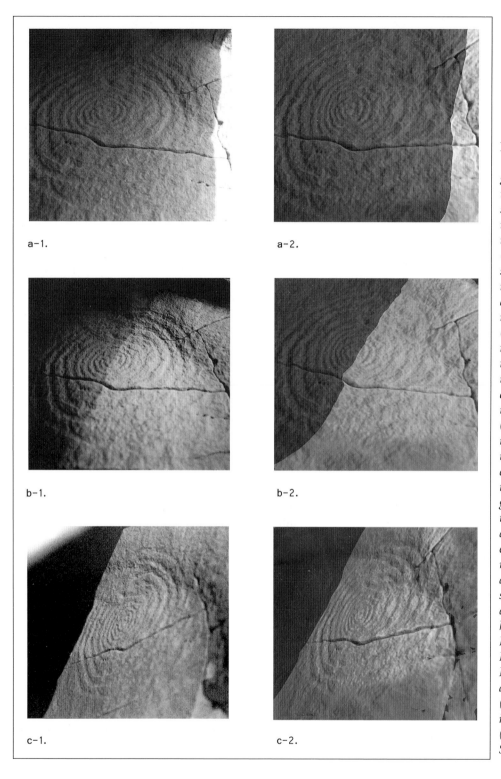

a-1.

a-2.

b-1.

b-2.

c-1.

c-2.

FIGURE 3. *Pairs of 1980 to 1987 photographs with 2006 registered model simulated images of rising sun and moon markings: **a**, equinox (September 23, 1980, sun rise at approximate azimuth 90.2 degrees); **b**, northern minor lunar standstill (May 13, 1980, using the sunrise to simulate the moon at approximate azimuth 67.4. degrees); **c**, northern major lunar standstill (November 8, 1987, moon rise at approximate azimuth 55.4 degrees). In each pair, the dates of the photographs are used in the simulations. The altitudes of the high edge of the disc of the rising sun or moon are taken in the simulations as 0.35 degrees above the true horizon of 0.2 degrees. Photos by Karl Kernberger (1978), Nevada Weir (1980) and Rolf Sinclair (1987) and simulated images by Alan Price (2006). (Copyright © Solstice Project.)*

spiral; and the small equinox light dagger did not pierce the smaller spiral (Palca 1989; Sofaer and Sinclair 1990).

The National Park Service response was two-fold: It limited visitation to the site more severely and brought in fill to replace the eroded sediments around the base of the slabs, constructing a coursed masonry wall east of the eastern most slab to prevent additional sediment loss downslope.

Computer Graphics Efforts.

Throughout the 1980s and the 1990s the Solstice Project pursued efforts to replicate the site in consultation with experts in the burgeoning field of computer graphics, turning from 1979 to 1981 to Massachusetts Institute of Technology's Computer Graphics Department and from 1983 to 1987 to the Math Department of Rensselaer Polytechnic Institute (RPI). A graduate student at RPI, Eric Brechner, developed a prototype interactive model of the site on a Silicon Graphics computer (Bordner 1989; Sofaer et al. 1989). In the early 1990s, a graduate student at Ohio State University's Center for Mapping, Ken Edmundson, with Phillip Tuwaletstiwa and Kurt Novak, developed a 3D model of a large portion of the site from the 1984 photogrammetric measurements. Kurt Novak combined Edmundson's digitized model with the interactive model from RPI to display on a Silicon Graphics computer a partial interactive model of the site (Novak et al. 1992).

While these efforts paved the way, limitations in both the nature of the measurements to date and in the available technology precluded the development of a fully and accurately functioning model of the Sun Dagger site. For example, the large photogrammetric cameras, which had to be stationary and at least forty centimeters apart, could not adequately record within the crevice between slabs and the spirals. This meant that Edmundson's model could replicate only the outer views the rock slabs, not the spirals or the inner edges of the slabs that cast half the shadow and light patterns that marked the solar cycle as well as the shadow patterns of the lunar cycle.

In 2001, Alan Price, then at the University of Maryland, and the Solstice Project created an interactive model of the Sun Dagger site based on the Edmundson model. Now on display at the Adler Planetarium and Astronomy Museum in Chicago, Price's model allows viewers to move around the site and to view light patterns created on the spirals at different times of the day and year. Yet, while educational and illustrative, the model is not an accurate or complete replication of the site. The Solstice Project's goal of developing such a model remained elusive.

DEVELOPING AND ORIENTING THE LASER SCAN MODEL OF THE SUN DAGGER SITE

In 2003, Western Mapping Company's James Holmlund proposed that with new laser scanning technology, his group could measure the inner edges of the slabs and the spirals, as well as the outer shapes of the slabs and the entire nearby cliff face and ground. Holmlund estimated the results would be accurate within one centimeter, a level of precision at the low end of the range the Solstice Project had estimated would be required to authentically replicate the light markings.[7]

The assumption was that by now all three slabs could have moved from their 1984 positions. The challenge concerned the plausibility of integrating two key sets of measurements: the accurate but limited and low sampling density of the 1984 photogrammetric data gathered before the site was disturbed, and the much more comprehensive and highly accurate laser scanning measurements to come. The conclusion: The 1984 photogrammetric record had sufficient three dimensional definition and accuracy—and, most critically, included enough tie-in with the cliff face—to allow a tight registration of the slabs of the proposed laser-scanned model with the slabs prior to their disturbance. Holmlund suggested that after acquiring three dimensional laser scan data for the current site configuration, the computer program could digitally separate the new slab models into their component parts and virtually reposition them to fit the configuration from the 1984 close-range terrestrial photogrammetric model.

With this assurance, the Solstice Project contracted with the Western Mapping Company to conduct the laser scanning of the Sun Dagger site. Critical technical support came from the National Park Service, which wanted an archival record of the site in its current condition as a base for monitoring future changes. Following extensive logistical planning with NPS archaeologist Dabney Ford, the scanning effort proceeded with her generous assistance and that of professional climber Scott Sholes, several volunteers, and four members of the NPS ruins stabilization crew. A crew of thirteen hauled 670 pounds of equipment to the site, and two Western Mapping staff members, Joseph Nicoli and William Haas, conducted the laser scanning and surveying during the afternoon and night of May 11 and the morning of May 12, 2005.

Accompanying the group were William Stone of the National Geodetic Survey of the National Oceanic and Atmospheric Administration (NOAA), who would provide precise geodetic coordinates for the laser-scanned model, and Alan Price, who would conduct, with Stone and Haas, accurately timed photo documentation of the light patterns at this time of the laser scanning. Price, who would develop the final interactive computer model, would first use this photo documentation of May 11-12 to test the accuracy of the laser model with simulations of the light patterns on these dates.

Establishing Geodetic Orientation.

On May 11 and 12, 2005, NOAA's William Stone positioned a permanent geodetic control point near the slabs. He established the geodetic position of the control point by collecting seventeen hours of static, dual frequency Global Positioning System (GPS) observations with survey grade GPS equipment and procedures. The National Geodetic Survey's Online Positioning User Service (OPUS) utility processed the resulting data with respect to the nationwide network of permanent GPS Continuously Operating Reference Stations

(CORS), which defines the nation's modern National Spatial Reference System. Peak-to-peak errors (that is, the approximate uncertainty of the results, with respect to the CORS network) of about one centimeter horizontally and two centimeters vertically characterized the resulting geodetic position.

Additionally, between May 11 and May 13, Stone collected GPS data on several vertical control points around Chaco Canyon to assess the accuracy and behavior of the GEOID03 model, which is used to convert between the ellipsoid height system (referenced to a mathematically defined surface, called the ellipsoid) used by GPS and the orthometric height system (referenced to a gravitationally defined surface, called the geoid) used on traditional maps to describe ground elevation. These additional observations indicated that GEOID03 works well in the Chaco area and that the resulting elevations derived from the combination of GPS measurements and the GEOID03 model are sufficiently accurate for this application.

Two sites provided azimuth control for the laser scan work: a recently established permanent GPS installation on the mesa top north of Fajada Butte and a Public Land Survey System section corner that was positioned with GPS during the project. The GPS installation's monument and the section corner both provide azimuth control for optical survey instrumentation (e.g., total station or theodolite) located at the Fajada Butte control point.

Verification of the May 11–12 results came with an independent survey by Stone of the Fajada Butte control point December 16–17, 2005. The findings of a sixteen hour GPS observation session, again processed through the OPUS utility, agreed with the earlier results within 0.5 centimeters in horizontal position and 1 centimeter in height.

Conducting the Laser Scanning.

The Western Mapping staff carried out the laser scanning in two phases:

In the first phase, the staff scanned the overall cliff face and most of the slab geometry with a Leica HDS 2500, which uses time-of-flight scanning.

With this instrument, a laser scanner with a known relative orientation emits a laser pulse; the time the pulse takes to return to the scanner establishes a point at the measured distance and at the orientation of that pulse. The scanner runs through this process at a rate of a thousand times per second, creating a point cloud. The scanner is then moved to another position, where the process is repeated. Setting the instrument in various locations allowed the Western Mapping staff to capture complex shapes, with intricate relationships, in three dimensions. The cliff face was scanned at a density of points of one centimeter (that is, a point was collected every centimeter) and the slabs were scanned at a resolution of five millimeters. The scanner recorded a total of 7.2 million points from thirty-one different scanner setups.

In the second phase, the Western Mapping staff conducted small scale triangulation scanning of the two spiral petroglyphs and most of the shadow casting edges of the slabs. The triangulation scanner consists of a camera and a line laser. The laser passes over the area at a known rate, while the camera captures 3D coordinates, triangulating between the camera, the laser, and the rock surfaces. Western Mapping staff conducted seventy-one triangulation scans of the scene, all with an estimated accuracy of better than one millimeter. (Because the pixel count of the camera determines the resolution of the final image, resolution cannot be precisely quantified.)

Registration of the Laser Scans.

The Western Mapping staff completed the registration in three phases:

In the first phase, the time-of-flight point clouds were registered locally to each other. Targets placed in the scanned scene act as 3D reference points to which overlapping clouds can be registered. In addition to the targets, a process called cloud-to-cloud registration uses similar geometry in overlapping scans to align two scans to each other. The computer operator tags two point clouds that roughly identify areas of overlapping geometry. Software algorithms then create a best fit alignment of the two clouds. Repeated multiple times between various clouds, this process creates an overall, best fit alignment. Incorporating the targets and the cloud-to-cloud alignments, the final local registration is frozen.

In the second phase of registration of the laser scans, the Western Mapping staff put the locally registered scans into a geodetic coordinate system so that the scan information could be modeled later with respect to astronomic relationships. Using the geodetic positions established by William Stone for the new NGS monument (station FAJADA) on Fajada Butte, the staff computed a geodetic azimuth and Universal Transverse Mercator (UTM) azimuth between FAJADA and the CORS station previously established at Chaco Canyon. (The Universal Transverse Mercator is a universally used x-y planar mapping coordinate system, which is rigorously related to the geodetic latitude-longitude system.) They checked these orientations against the position of another GPS positioned section corner monument more than 1.5 kilometers from the site. Using a certified geodetic quality total station, the staff mapped the positions of reflective targets in the scan scenes and established world (UTM) coordinates for each target. These coordinates were then adjusted (see the section *Positioning and Orienting the Laser Scanned Model*) and incorporated into the cloud registration, moving the local registration into UTM based coordinates.

In the third phase, the Western Mapping staff used a different software program to group the small scale scans into five areas and register them to each other using the cloud-to-cloud method just described. The world coordinate registered group was then imported into the new software, where the overall cloud-to-cloud registration was further refined. Again using the cloud-to-cloud method, the small scale scan groups were then registered to the world coordinate group.

Modeling of the 2005 Laser-Scanned Site.

To model the laser-scanned site, the Western Mapping staff moved both groups of scanned data—the time-of-flight and the small scale scans—into a different software program. With the time-of-flight point data reduced in resolution to eliminate overlapping data points, a Triangular Irregular Network (TIN) was created at approximately six millimeters resolution. Five additional small scale scanner models were each decimated to 0.8 mm resolution and meshed independently. The resulting high resolution models were pasted into the overall model. These high resolution portions simply replaced less resolved sections (at 6 mm resolution) modeled from the time-of-flight data. The models were cleaned of spurious data and resampled using curvature based algorithms.

Positioning and Orienting the Laser-Scanned Model.

Because the survey was adjusted to the UTM coordinate system, an additional rotation was necessary to establish the geodetic orientation to match the input required for Alan Price's astronomical modeling program. The Western Mapping staff used the convergence angle between the UTM coordinates and geodetic positions computed from the reduction of the GPS data at station FAJADA to rotate the orientation of the model to geodetic north. Since the coordinates were originally UTM grid coordinates, they used the computed ground scale factor for FAJADA to scale the coordinates to enable accurate (local) model measurements. At this juncture, the coordinates of this adjusted (Sun Dagger site) system were no longer UTM coordinates, but a local coordinate system.

Testing the Accuracy of the Laser-Scanned Model.

To test the new model, Alan Price used the accurately timed digital photographs and slides of the light and shadow formations on the spirals and the surrounding cliff face taken during the two days of laser scanning—two sequences of images made during midday on May 11 and 12, and a third of sunrise on May 12. Independent of the fact that the light patterns have changed since the site's initial documentation in 1978, it was critical that simulations of light and shadow casting on the new digital model exactly match the site in its May 11-12, 2005 state.

To facilitate testing of the model before developing the interactive application, Price selected Alias Maya software because of its suitability for handling the high resolution model, for ray trace rendering, and for the custom scripting ability of Maya Embedded Language (MEL). In Maya, a directional light model was used to project orthogonal (parallel) light to ray trace shadow patterns from the slabs to the cliff face based on ephemeris calculations of the sun and moon positions. (An ephemeris is a table of the positions of a celestial body at regular intervals.)

To calculate the ephemeris, a series of MEL scripts was created that interfaced with an external application for the calculations. Code was compiled from Steve Moshier's AA code (www.moshier.net) for computing ephemerides of sun and moon using rigorous reduction methods from the *Astronomical Almanac* and related sources, and a long term extension of modern lunar theory for the moon's position. The results of the calculations were compared with the U.S. Naval Observatory ephemeris calculations and with a number of commercial planetarium software programs, including Starry Night Pro, Software Bisque's The Sky, and Sky Map Pro. The UTM coordinates and elevation obtained during the process of scanning the site were used as input parameters for the ephemeris calculations. A series of renderings was created with the digital model to correspond with the timing of images taken May 11–12, 2005.

The first results proved a highly accurate match of light patterns between the simulation and the photographic documentation, indicating that the processes of scanning, gathering positional and orientation data by the Western Mapping Company and William Stone of NGS, and Western Mapping's conversion process of the digital model had created an accurate digital reproduction of the site as it existed in May 11-12, 2005.

DEVELOPING THE 1984 PHOTOGRAMMETRY COMPUTER MODEL OF THE SUN DAGGER SITE

In the spring of 2006, Aero-Metric, under the direction of Andrew Piscitello, conducted further readings of the 1984 photogrammetry, eleven stereo pairs of glass plates recorded by Dennett, Muessig, and Ryan. A special effort was made to include as many data points as possible in this reading that would accurately relate the slabs to the cliff face, on the assumption that all three slabs might have moved by the time of Western Mapping's 2005 laser scanning project. The cliff provided the only stable feature of the site for both the 1984 and 2005 models of the slabs.

Background: 1984 Photogrammetric Activity.

In 1984, Dennett, Muessig, and Ryan had acquired eleven stereo pairs of photography of the Sun Dagger site on glass plates using a WILD 40 dual camera system. This system is comprised of two calibrated metric cameras mounted on a fixed base bar to ensure that their optical axes are parallel. Positioned a proper distance from an object, the system records overlapping images suitable for stereoscopic viewing. Using a stereoscopic restitution instrument to record x, y, and z positions of any point imaged in the object space, precise measurement is then possible from the stereoscopic images.

Dennett, Muessig, and Ryan had also conducted a field survey to determine the x, y, and z positions of control targets — small concentric rings (a bulls eye) affixed to the surface area of interest — which would serve to reestablish the position of the cameras during the photogrammetric restitution process. The object of the surveys, completed in a local coordinate system with Polaris observation (to determine true north), was to tie the observations to previous surveys by Koogle and Pouls that had located the site in geodetic position. The surveyors used a WILD T2 theodolite to attain their readings. The field surveys were designed to produce accuracies in the range of two millimeters.

Dennett, Muessig, and Ryan contracted with Aero-Metric to reduce the fieldwork surveys and complete the photogrammetric restitution. Aero-Metric reduced the survey work in the local coordinate system with orientation to true north but did not transform the results to the actual geodetic location because the necessary data were not available. In 1984, Aero-Metric completed an analytic triangulation of all the photographs and produced a simultaneous adjustment of all the photo positions to verify the control points. The photogrammetric restitution of each of the eleven stereo models was then completed on a WILD BC1 analytical stereo plotter, and several thousand points on the surface of the Sun Dagger slabs and on the large and small spiral petroglyphs were recorded. The photogrammetric accuracies varied from a couple of millimeters to approximately ten or twelve millimeters, according to the variation of the distance of the cameras from the image points. This effort did not produce a complete computer model of the Sun Dagger site because the data set was not sufficiently dense and the system was unable to measure points within the narrow openings between and behind the slabs.

2006 Photogrammetric Reading.

In the spring of 2006, at the request of the Solstice Project , Aero-Metric used the 1984 photogrammetry survey to add points on the cliff wall, against which the slabs were positioned, to its original readings of this data set. The stable cliff wall and its features were accurately shown in the LiDAR data set. In particular, Aero-Metric added readings from the 1984 data set of micro topographic features of the cliff wall so that the photogrammetric model could be accurately transformed to the 2005 laser scanned model.

For this new work, the individual stereo models were reset on a Zeiss P1 analytical stereo plotter. Once the existing data set was verified, Aero-Metric added new recordings of x, y, and z coordinates of the significant features on the cliff wall that the Western Mapping Company had identified, based on the laser scanned data, as criti-

159

cal areas for registering the two models. These features could readily be identified in the LiDAR data set. The resulting new and more comprehensive (1984) photogrammetric model was transferred to the Western Mapping Company in Tucson.

REGISTERING THE 1984 PHOTOGRAMMETRIC MODEL WITH THE 2005 LASER SCAN MODEL

Western Mapping Company staff digitally removed the disturbed slabs in the 2005 laser scanned model, maintaining only the laser scanned cliff face. They then registered the slabs and the cliff face of the new 1984 photogrammetry model to the laser scanned cliff face. Using cloud-to-cloud registration, they then registered each of the laser scanned slab models that had been digitally removed to its corresponding, premovement version in the new model. The resulting registration was then frozen as the 1984 model.

Beyond an attempt to digitally restore the slabs to their 1984 positions, this registered model contained remarkable detail of the site, with more than thirty million points, compared to the nine thousand points of the 1984 photogrammetry. (It should be noted that the significantly smaller number of points in the 1984 model did not detract from its high accuracy in modeling the slabs and their relationship to the cliff face, because the points were collected in smaller critical areas of the slabs and bedrock that had been identified in the laser model.) The laser scanned model covered far more of the site than did the 1984 photogrammetry, encompassing the surrounding cliff and ground areas up to five meters above the slabs, three meters to the east and three meters to the west. For example, it recorded a critical shadow casting edge located approximately 4.7 meters above the slabs, in an area suspected, and later specifically identified, as defining the upper edges of the light daggers in the summer season.

In addition, the new model made possible accurate geodetic orientation, which the 1984 photogrammetric survey had not obtained, derived from the 2005 geodetic positioning of the FAJADA monument at the site by NGS.

TESTING THE ACCURACY OF THE REGISTERED MODEL

To test the accuracy of the registered model from Western Mapping, Alan Price in the spring of 2006 applied the astronomical program that he had used in his test of the 2005 laser model. As an integration of the detailed 2005 laser scanned model into the 1984 photogrammetric model, the registered model should have restored the slabs to their positions prior to their disturbance and thereby simulated images of the light patterns, replicating the early photo documentation of the site.

Price's test results showed that the solar and lunar images simulated by the registered model match the Solstice Project's accurately timed photo documentation of solar and lunar events at the site in the years 1978–1987. (In addition to the set of sequential photographs taken by Karl Kernberger, numerous other timed photographs had been taken of lunar and solar events at the site.) The paired images in Figures 2 and 3 show light and shadow markings photographed between 1978 and 1983 at the key times of the solar and lunar cycles compared with those simulated by the registered model at these times.

The Solstice Project's first goal was now a reality: an archival digital replication of the Sun Dagger and its astronomical functioning.

ASSESSING THE SITE'S DISTURBANCE

Using the two models—the 2005 laser scanned model and the 1984 photogrammetric model—Price and Western Mapping explored the extent of disturbance to the slabs. Figure 4 illustrates Price's comparison of the simulated slabs restored to their 1984 positions, shown in solid form, and the 2005 disturbed slabs, shown in black outline. The middle slab had moved the most of the three—15 centimeters on one axis. The other two slabs had also moved: the eastern slab, 5.4 centimeters on a similar axis, and the western slab the least, 0.8 centimeters.

FIGURE 4. *The slabs in an image developed from the registered model and the laser-scanned model that shows their original positions, outlines of their disturbed (2005) positions, and measurements of certain of the differences between the 2005 and 1984 positions. Simulated image by Alan Price. (Copyright © Solstice Project.)*

EXPERIMENTS WITH THE GEOLOGICAL HISTORY OF THE SUN DAGGER SITE

Preliminary experiments with the 2005 registered model have placed the slabs in the positions where, according to geological reports (Sofaer *et al.* 1979; Newman *et al.* 1982), they were originally attached to the nearby cliff along an approximate horizontal plane. Plans call for more detailed efforts to match the geometry of the cliff face with the inner surfaces of the slabs to verify or refute these earlier reports, and possibly to further pinpoint the slabs' place of origin. With the slabs attached to the cliff in the computer model, planned kinetic experiments should show possible positions to which the slabs could have fallen and been found by the Chacoans.[8]

CREATING AN INTERACTIVE RESEARCH MODEL

In 2006, Alan Price created an interactive application for research and analysis of the site. The application combines functionality similar to the tools created with MEL scripting with new navigation and manipulation tools, so that researchers can explore the model in a stand-alone application.

The interactive application allows one to navigate around the 3D model, observing it from any angle. One can set the calendar date and time of day for positioning the sun and moon, projecting the shadows of the stone slabs onto the cliff and spiral patterns in real time. Translation and rotation tools allow one to experiment with manipulation of the slabs and/or the cliff face in the region of the spirals, as well as to alter the latitude, longitude, and orientation of the entire site. One can set markers at any position on the surface of the slabs or cliff face to measure distances between points and to determine shadow casting edges. A surface modeling tool allows one to alter the shape of the slabs, much as one might use a small tool to work the shadow casting edges. One can alter the orientation and size of the spiral patterns or "draw" an entirely new spiral. These tools allow a user to deconstruct the site, experimenting with variations to gain a better understanding of the complexity and precision with which it operates.

CONCLUSION

Through the extraordinarily generous and dedicated efforts of many individuals, bringing rich interdisciplinary experience to the project, the Chacoan Sun Dagger is digitally restored. It is now also accessible for challenging research explorations and, it is hoped, will serve as a stimulating educational resource.

Centuries ago, the Sun Dagger site engaged skilled and trained astronomers to achieve its complex astronomical expressions. It is of interest that its restoration required several of today's most advanced technologies, employed by scientists with modern math and engineering backgrounds.

Several mapping specialists of Western Mapping Company and a geodesist ensured remarkable precision and scope in the laser model's replication of the site and its orientation. Photogrammetrists provided high quality stereo glass plates of the site before its disturbance. In the past year they also provided the thorough readings of these plates for the intricate detail that was *essential* for the registration of the recent laser model to the model of the site prior to its disturbance. Recently developed computer software allowed the registration of the two models. Finally, a computer modeler, using the latest programs of astronomical data and sophisticated interactive computer applications, could test the model of the restored slabs against the 1978 accurately timed photo documentation of the astronomical markings of the site. The success of the digitally restored model is evident in its exact replication of this early record, with no manipulation of any element in the model.

In contrast, about a thousand years ago, the Chacoans developed the site by applying their knowledge of the solar and lunar cycles and their astute observations of shadow and light patterns to three sandstone slabs located near a south-southeast-facing cliff face. As in the alignments of their buildings (see note 4), the Chacoans developed in the light markings of the Sun Dagger site an integrated expression of the sun and moon. This site and the Chacoans' other elaborate astronomical works are physical realizations of cosmological concepts.

The detail and precision of the research model, with its numerous interactive tools, offers opportunities to analyze the Chacoans' process.[9] Experimentation with the model should bring insights about the knowledge, planning, and experimentation the Chacoans employed to achieve the interlocking markings of the sun and moon. Already it has revealed, to one researcher, elements of the site that were readily available and suggestive to the Chacoan astronomers. Further exploration may reveal the Chacoans' process of refining their light markings by shaping, moving, or adjusting elements of the site.

Perhaps, paradoxically, only an interactive research model achieved with the latest technologies will allow modern researchers and students to appreciate the Chacoans' capacity to conceptualize and work, *without* such technology, with the four dimensions of time and space as they created the Sun Dagger site.

ACKNOWLEDGMENTS

We appreciate the generous help to this project by the Chaco Culture National Historic Park. In particular, we appreciate the thoughtful planning, safety preparation, and logistical support by park archaeologist Dabney Ford and the NPS ruins-stabilization crew. Rolf Sinclair and Karl Kernberger contributed generously to creating the valuable timed photodocumentation of the Sun Dagger site prior to its disturbance. The National Geodetic Survey of NOAA assured us the assistance again of William Stone in producing a superbly precise geodetic survey of the site. We thank Phillip Tuwaletstiwa for his early and continuing interest in the Sun Dagger modeling effort, and Kenneth Edmundson for his dedicated work to create the early photogrammetric model of the site. In 1984, Dennett, Muessig, and Ryan recorded the precise photogrammetric stereo pairs used in the final modeling. The Center for Mapping of Ohio State University, the Imaging Research Center of the University of Maryland, ACCAD of Ohio State University, Aero-Metric, Inc. and Cooper Aerial, Inc., provided critical technical assistance to our modeling efforts. We thank NPS archaeologist Roger Moore, and volunteers Scott Sholes (Durango, Colo.), Susan Yewell (Solstice Project), Craig Johnson (Santa Fe), Armando Espinosa Prieto (Santa Fe), and Richard Friedman (Farmington, N.M.) for their generous logistical and safety support to our laser scanning of the Sun Dagger site in May 2005. Lindsay Kraybill (Western Mapping Company) provided critical data analysis in developing the 2005 laser model and the registered model.

NOTES

1. See the references in the first paragraph of this paper for discussions of possible cultural affiliations of the Sun Dagger site, estimating the time of its development between A.D. 900 and 1300 and for certain parallels between the Chacoan astronomical expressions and the traditions of the historic Pueblo cultures.

2. The Solstice Project *(www.solsticeproject.org)* is a nonprofit organization dedicated to the study of ancient cultures of the American Southwest, founded in 1978 by Anna Sofaer to study, document, and preserve the Sun Dagger and other astronomical expressions of these cultures.

3. It is difficult to strictly define at what point the shadow and light formations of the rising major and minor standstill moon and the equinox sun can be considered "markings." These formations appear, however, to be distinctive patterns (i.e., close to the edges and the center of the large spiral) when they are cast by the sun or moon rising within 0.5 to 1.5 degrees of the values given here of azimuths 54.3, 67.1, and 90 degrees (see Sinclair and Sofaer 1993 for how these values were established). In our documentation, we define "rising" as when the sun or moon is 0.5 to 1.5 degrees above the 0.3-degree altitude of the eastern-north-eastern horizon. See annotations to Figures 2 and 3 for the specific azimuths of the solstice, equinox and lunar standstill light and shadow markings that are illustrated in this paper.

4. Following the discovery of these astronomical markings at the Sun Dagger site, research by the Solstice Project showed that twelve major Chacoan buildings in and near Chaco Canyon are aligned to the extremes or midpositions of the solar and lunar cycles (Sofaer 1998; Sofaer in press). A non-Project researcher documented the relationship of Chimney Rock, a Chacoan building in southeastern Colorado, to the rising of the northern major standstill moon (Malville and Putnam 1989). In addition, fourteen major Chacoan buildings incorporate an internal solar-lunar geometry and relate to each other over distances on alignments to the sun and the moon (Sofaer 1998; Sofaer in press 2007).

5. Note that the right and left edges of the large spiral are marked multiply: at the winter solstice with two light daggers on these edges; at the equinox by the rising sun's shadow on the right edge; and at the major lunar standstill by the rising moon's shadow on the left edge. In addition, the minor standstill moon's shadow and the summer solstice sun's light dagger both mark the center of the spiral. Finally, the summer solstice dagger also marks the top turn of the spiral: for the four to five days before and after summer solstice only the light form first appears as a spot of light in the top turn. A week after summer solstice, a streak of light appears above the spiral and expands through the seasons. In sum, these numerous interlocked markings appear to 'define' the shape and size of the large spiral. (For a fuller discussion of this apparent coordination of the markings at the site, see Sinclair and Sofaer 1987).

6. NPS records showed that more than a thousand registered visitors had been to the site between 1977 and 1982. Many others, it is assumed, visited the site without registering with park staff (Trott *et al.* 1989). See Figures 3b (1) and (2). The loss of material along the shadow casting edge of the eastern slab caused the difference evident between the image of the 1980 photograph 3b (1), taken before the loss of this material, and the simulated image of the same, derived from the laser scan that registered the change 3b (2). See in particular the difference in the shadow line below the spiral center.

7. As it turned out, the model has only about half the inaccuracy Holmlund estimated.

8. In 1982, three authors reported their geological and archaeological assessment that the slabs *could* have fallen into their 1978 recorded positions (Newman *et al.* 1982), refuting an earlier report of the Solstice Project (Sofaer *et al.* 1979). This report strongly implied, by its title and concluding statements, that there had been no movement of the slabs by the Chacoans, although it provided no data excluding this possibility. Following this proposal, two of the authors reported that the slabs could have been adjusted and shaped to achieve the markings (Simon 1982). See Sofaer and Sinclair 1987 for discussion of the likelihood of the Chacoans achieving the complexity of the solar and lunar markings at the site without some adjustment to the slabs.

9. Ben Luce, a theoretical physicist, has recently initiated a study of the Sun Dagger site's functioning, using the interactive tools of the computer graphics model and other theoretical models based on the interactive model, to explore the mechanisms involved and possible constructive aspects. He reports that preliminary research "reveals a fully but not overly determined system in the way the site controls a complex set of astronomical and geological variables to achieve its markings."

REFERENCES

Bordner, K.
1989. "Computer Graphics Unlock Mysteries of the Past." *Rensselaer*, March, 4-7.

Malville, J. M., and C. Putnam.
1989. *Prehistoric Astronomy in the Southwest*. Johnson Books, Boulder, Colo.

Moshier, S.
"Astronomy and Numerical Software Source Codes." *www.moshier.net*.

Newman, E. B., R. K. Mark, and R. G. Vivian
1982. "Anasazi Solstice Marker: The Use of a Natural Rockfall." *Science* 217: 1036–1038.

Nicoli, J., J. Holmlund, A. Sofaer, A. Price, L. Kraybill, and W. Stone
2006. "Digital Restoration of the Sun Dagger Site, Chaco Canyon." Abstract, Society for American Archaeology, San Juan, Puerto Rico.

Novak, K., K. L. Edmundson, and P. Johnson
1992. "Spatial Reconstruction and Modeling of the Sundagger Site in Chaco Canyon." *ISPRS International Archives of Photogrammetry and Remote Sensing* 29, Commission 5, Part B5: 808–812.

Palca, J.
1989. "Sun Dagger Misses Its Mark." *Science* 244:1538.

Simon, C.
1982. "Solar Marker: A Natural Rock Fall?" *Science News* 122, No.19: 300

Sinclair, R. M., and A. Sofaer
1993. "A Method for Determining Limits on the Accuracy of Naked-Eye Locations of Astronomical Events," in *Archaeoastronomy in the 1990s,* ed. Clive Ruggles, 178–184. Group D Publications, Loughborough, U.K.

Sinclair, R. M., A. Sofaer, J. J. McCann, and J. J. McCann Jr.
1987. "Marking of Lunar Major Standstill at the Three-Slab Site on Fajada Butte." *Bulletin of the American Astronomical Society* 19: 1043.

Sofaer, A.
1998. "The Primary Architecture of the Chacoan Culture: A Cosmological Expression," in *Anasazi Architecture and American Design,* ed. B. Morrow and V. B. Price, University of New Mexico Press, Albuquerque, 88-132; also in press 2007 in *Architecture of Chaco Canyon, New Mexico,* ed. S. H. Lekson. University of Utah, Salt Lake City.
1982. Protective Measures for the Sun Dagger Site, Chaco Canyon National Historic Park, Fiscal Year 1983. Memorandum to Senator Pete V. Domenici.

Sofaer, A., and R. M. Sinclair
1990. "Changes in Solstice Marking at the Three-Slab Site, New Mexico, U.S.A.," *Archaeoastronomy,* no. 15 (JHA, xxi) 59-60.
1987. "Astronomical Markings at Three Sites on Fajada Butte, Chaco Canyon, New Mexico," in *Astronomy and Ceremony in the Prehistoric Southwest,* ed. J. Carlson and W. J. Judge Jr., 13–70. Maxwell Museum of Anthropology, Albuquerque.

Sofaer, A., R. M. Sinclair, and E. Brechner
1989. "Computer Graphics Model of the Three-Slab Site on Fajada Butte, New Mexico." *Bulletin of the American Astronomical Society* 21: 1210.

Sofaer, A., R. M. Sinclair, and A. Doggett
1982. "Lunar Markings on Fajada Butte, Chaco Canyon, New Mexico." In *Archaeoastronomy in the New World,* ed. A. F. Aveni, 169–181. Cambridge University Press, Cambridge, U.K.

Sofaer, A., V. Zinser, and R. M. Sinclair
1979. "A Unique Solar Marking Construct." *Science* 206: 283–291.

Trott, J., R. K. Mark, D. B. Fenn, and C. Werito
1989. *Evaluation of Status of Site 29 SJ 2387: Chaco Culture National Historical Park.* National Park Service.

U.S. Congress. 1981. *Congressional Record,* December 9, S14283.

Appendix:
Abstracts

A.

MARKING OF LUNAR MAJOR STANDSTILL AT THE THREE-SLAB SITE ON FAJADA BUTTE

Rolf M. Sinclair (NSF), Anna Sofaer (Solstice Project),
John J. McCann (Polaroid Corp.), and John J. McCann, Jr. (Belmont H.S.)

Bulletin of the American Astronomical Society, 171st Meeting Abstracts, Vol. 19, No. 4, 1987, p. 1043

At this site in Chaco Canyon, New Mexico,[1] a carved spiral is partly shielded from sunrise (or moonrise) by one of three stone slabs. The edge of the shadow cast by this slab on the spiral is a sensitive indicator of the azimuth (hence declination) of the rising sun or moon. We had earlier[2] found that at declinations of zero and 18.4° this shadow edge falls tangent to the right edge and through the center of the spiral respectively, and had predicted that at declination 28.7° (the lunar major standstill) the shadow would be tangent to the left edge. Direct observation of the shadow cast at moonrise at maximum declination in 1987 (the year of the major standstill) shows that this in fact occurs. This site thus singles out the northern extreme declinations of the lunar 18.6 year cycle (major and minor standstills) as well as the midpoint of the lunar (and solar) motions.

1. A. Sofaer *et al., Science,* 206, 283-291 (1979).

2. A. Sofaer *et al.,* in *Archaeoastronomy in the New World* (A.F. Aveni, ed.),
   pp. 169-181. Cambridge University Press, (1982).

B.

A REGIONAL PATTERN IN THE ARCHITECTURE OF THE CHACO CULTURE OF NEW MEXICO AND ITS ASTRONOMICAL IMPLICATIONS

A. Sofaer (Solstice Project), R.M. Sinclair (NSF), and R. Williams (RPI)

Bulletin of the American Astronomical Society, 171st Meeting Abstracts, Vol. 19, No. 4, 1987, p. 1044

The orientations of external walls and the locations of 17 of 18 major Pueblo structures of the Chaco culture of New Mexico indicate that these structures are organized in a complex regional pattern. Earlier work showed the use of cardinals in major axes and external walls of Pueblo Bonito and Casa Rinconada (R. Williamson in *Archaeoastronomy in the Americas,* Ballena, 1981) and a cardinal bearing between Pueblo Alto and Tsin Kletzin (J.M. Fritz in *Social Archaeology,* Academic Press, 1978). We find that the orientations of two pueblos—Pueblo Alto and Hungo Pavi—are also cardinal and that the interpueblo bearing between Casa Rinconada and New Alto is cardinal. We observe the repetitious use of certain other orientations in external walls and their interpueblo bearings. The interpueblo bearings link pueblos over distances ranging up 55 km. In some cases these bearings link walls that share the same orientation. These pairs of linked walls have orientations which differ by only 0.25° and 1.0°. The orientation of the external walls of two or more pueblos may link with a single pueblo and the orientation of the external walls of a single pueblo may link with two or more pueblos. Mapping these interpueblo bearings reveals a complex, symmetric interrelationship between the 17 major structures in a 5,000 sq. km area. Several of the Chaco roads repeat the interpueblo bearings and are associated with pueblos linked into the pattern by these bearings. The regional pattern presented here offers the first explanation for the specific location and orientation of many of Chaco's major structures. The orientations of many of the external walls and interpueblo bearings align with the lunar standstill positions. We discuss whether or not these possible lunar alignments are intentional or merely coincidental. We note that the rigorous repetition of specific bearings in pueblo walls and locations could only be achieved with some method of astronomical orientation—perhaps one based in part on solar and shadow observation. We explore the parallels between general concepts of the historic pueblo cosmology and the patterning evident in the prehistoric constructions of Chaco.

C.

SOLAR SIGNIFICANCE OF A DOUBLE SPIRAL PETROGLYPH IN CHACO CANYON, NEW MEXICO

Anna Sofaer (Solstice Project), Rolf M. Sinclair (NSF), and Eric Brechner (RPI)

Bulletin of the American Astronomical Society, 173rd Meeting Abstracts, Vol. 20, No. 4, 1988, p. 991

We present evidence that a double spiral petroglyph in Chaco Canyon, New Mexico, the center of a prehistoric Pueblo culture, may depict the integration of the daily and seasonal cycles of the sun. This petroglyph is positioned on a vertical cliff of Fajada Butte so that the shadows cast by nearby cliff edges form a spear of light that bisects one of its linked spirals just at noon on equinox (A. Sofaer and R. M. Sinclair, *Astronomy and Ceremony in the Prehistoric Southwest,* J. Carlson & W.J. Judge, ed., Maxwell Museum, 1987). A double spiral can be generated by recording on each day from winter to summer solstice, the arc traced on a horizontal surface by the shadow of a fixed point (such as the tip of a stick) during a major part of the daylight hours, and connecting these successive daily arcs in a smooth curve (A. Sofaer, Archaeoastronomy, Vol. 3, No. 1, 1980; C. Ross, *Solar Convergence Solar Burn,* University of Utah Press, 1976). The light marking on the petroglyph corresponds with the identification of this particular double spiral as a depiction of the sun's cycles. It is significant that a configuration that can be derived from patterns of shadow and light created by the motions of the sun itself should be marked by a bisecting pattern of shadow and light just at midpoints of the sun's daily and seasonal cycles.

Fajada Butte displays abundant evidence of other uses by the Chaco culture of shadow and light patterns to mark the midpoints and extremes of the sun's daily and seasonal cycles (A. Sofaer *et al., Science,* 206, 1979; A. Sofaer and R.M. Sinclair, 1987, previous reference.). We will discuss these and other integrated and abstract expressions by the Chaco culture of the sun's cycles (A. Sofaer and R.M. Sinclair, *World Archaeoastronomy,* A. Aveni, ed., Cambridge University Press). We will also note references among the Mesoamerican cultures and the historic Pueblo people to the double spiral as a depiction of the solar and other celestial cycles.

D.

AN INTERPRETATION OF A UNIQUE PETROGLYPH IN CHACO CANYON, NEW MEXICO

Anna Sofaer, Rolf M. Sinclair

World Archaeoastronomy, edited by A. F. Aveni, Cambridge University Press, 1989, p. 449.

The design of a petroglyph on Fajada Butte in Chaco Canyon is a half circle surmounted by a spiral. This pattern corresponds closely to the ground plan of Pueblo Bonito, a massive semicircular structure that is the largest construction in the prehistoric U.S. southwest, and to this building's relationship to the solar cycles. The recumbent "D" shape of the petroglyph, unusual in Chacoan rock art, is the same shape as the outline of this pueblo. Two lines of the petroglyph (the diameter and the perpendicular radius) appear to represent two major walls of Pueblo Bonito that are accurately aligned north-south and east-west. In addition, the location of a drilled hole in the petroglyph corresponds to the location of the primary kiva of the pueblo. The spiral design above the "D" appears to represent the sun's daily and seasonal passage over the pueblo and to refer to the pueblo's cardinal alignments and solar orientation. The carving also shows the pueblo design as bow-and-arrow, a symbol associated with the sun in the mythology and ritual of the successor historic pueblos. The use of this petroglyph to express solar phenomena is consistent with the use of other petroglyphs near the top of the same butte, where sunlight forms a number of noon-seasonal markings at the solstices and equinoxes on spirals and other shapes. This recognition of the petroglyph as a solar symbol of Pueblo Bonito is the first reading of a petroglyph as an explicit statement of cosmology and architecture.

E.

COMPUTER GRAPHICS MODEL OF THE THREE-SLAB SITE ON FAJADA BUTTE, NEW MEXICO

Anna Sofaer (Solstice Project), R. M. Sinclair (NSF), and Eric Brechner (RPI)

Bulletin of the American Astronomical Society, 175th Meeting Abstracts, Vol. 21, No. 4, 1989, p. 1210

An interactive dynamic computer graphics model of the three-slab site on Fajada Butte, New Mexico will be demonstrated. The model archivally duplicates the astronomical functioning of this ancient Pueblo Indian site (c. A.D. 950-1150). The three slabs have been collimating sunlight and moonlight to form patterns on two spiral carvings on the cliff face. These patterns have marked the summer and winter solstices and the equinoxes (A. Sofaer, V. Zinser, and R.M. Sinclair, *Science,* 206, 283-291, 1979). In addition, one of the slabs casts shadows at the rising of the moon in patterns that mark the lunar standstill cycle (A. Sofaer, R.M. Sinclair and L. Doggett, *Archaeoastronomy in the New World,* A. Aveni, ed., 169-181, Cambridge University Press, 1982). A major change has been recently observed in the summer solstice marking of the site (Preliminary Report on Recent Change Observed at the Sun Dagger Site in Chaco Canyon, New Mexico, A. Sofaer and R.M. Sinclair, Memorandum, Aug. 1, 1989). The photogrammetry on which this model was based was recorded before this recently observed shift (S. Curry, D. Fair, D. Encinias, A. Sofaer and R.M. Sinclair, *Technical Papers 1987 ASPRS-ACSM Annual Convention:* Vol. 2. Photogrammetry, 1-7).

The model will demonstrate the degrees of sensitivity in the shapes and positioning of the slabs in relationship to the solar and lunar patterns, and it may also suggest the extent of natural and human activity in the development of the site's astronomical functioning. Certain of the techniques employed in this model may be useful to other scholars studying other ancient astronomical sites. This model was developed jointly by the Solstice Project and Rensselaer Polytechnic Institute, with photogrammetric surveying and digitization by Vexcel and Koogle & Pouls.

F.

DIGITAL RESTORATION OF THE SUN DAGGER SITE, CHACO CANYON

J. Nicoli,[1] J. Holmlund,[2] A. Sofaer,[3] A. Price,[4] L. Kraybill,[5] and W. Stone[6]

Paper presented at 72nd Annual Meeting of the Society for American Archaeology, San Juan, Puerto Rico, 2006.

The Sun Dagger Site, rediscovered in 1977, is an ancient Chacoan calendrical site consisting of three upright sandstone slabs casting light and shadow patterns onto two spiral petroglyphs marking solar and lunar cycles. In 1989, the Solstice Project found that the middle slab had pivoted from its original position. In 2005, using GPS and 3D LiDAR methods and legacy (1983) photogrammetric measurements, researchers produced two high-resolution 3D computer models of the 1983 and current (2005) positions of the slabs. The "restored" computer model will allow researchers to virtually study the complex astronomical relationships evident at the Sun Dagger site.

1. Western Mapping Company
2. Western Mapping Company
3. Solstice Project
4. Ohio State University
5. Western Mapping Company
6. National Geodetic Survey

About Anna Sofaer

Photo: Michael Pertschuk

ANNA SOFAER discovered the Sun Dagger site atop Fajada Butte in Chaco Canyon in 1977 while recording petroglyphs as a participant in a field school organized by the Archaeological Society of New Mexico. This finding initiated her continuing studies of this and other Chacoan astronomical sites and a three-decade engagement with archaeo-astronomy—the study and recording of astronomical observations and patterns by ancient peoples.

A 1962 graduate of Sarah Lawrence College, she had worked previously in social planning, community organization, housing opportunity initiatives, and the arts. While a working artist, she developed an interest in Maya astronomy and the pictographs and petroglyphs of the Southwest and became intrigued by implications that certain rock art might relate to astronomical observations.

Following her early encounters with the Fajada Butte site, Sofaer founded the Solstice Project in 1978 in Washington, D.C. She directs its activities today from Santa Fe, New Mexico. The non-profit Solstice Project conducts research, preservation and educational efforts on the astronomical expressions of the Chacoan culture of the Southwest. Sofaer coordinates and documents the interdisciplinary research of the Solstice Project, resulting in numerous scientific papers. This work has included extensive consultations and sharing of her findings with Pueblo educators and historians who regard Chaco as a sacred place with primary significance in their history.

In 1982 Sofaer produced, directed and co-wrote *The Sun Dagger,* an hour-length documentary narrated by Robert Redford and telecast nationally by the Public Broadcasting Service (PBS) and by the Discovery and A&E cable channels. In 2000 she produced, directed and co-wrote *The Mystery of Chaco Canyon,* an hour-length film also narrated by Redford and telecast nationally by PBS and by the National Geographic Society on its international cable network. Currently the Solstice Project is studying parallels and relationships of the Chacoan expressions of astronomy with those of Mesoamerican cultures.

Anna Sofaer and her husband Michael Pertschuk live and work today in Santa Fe.